Andreas Licht

The transcription factor CcpN from Bacillus subtilis

Andreas Licht

The transcription factor CcpN from Bacillus subtilis

An in-depth characterisation of a bacterial transcription regulator

Südwestdeutscher Verlag für Hochschulschriften

Impressum/Imprint (nur für Deutschland/ only for Germany)
Bibliografische Information der Deutschen Nationalbibliothek: Die Deutsche Nationalbibliothek verzeichnet diese Publikation in der Deutschen Nationalbibliografie; detaillierte bibliografische Daten sind im Internet über http://dnb.d-nb.de abrufbar.

Alle in diesem Buch genannten Marken und Produktnamen unterliegen warenzeichen-, marken- oder patentrechtlichem Schutz bzw. sind Warenzeichen oder eingetragene Warenzeichen der jeweiligen Inhaber. Die Wiedergabe von Marken, Produktnamen, Gebrauchsnamen, Handelsnamen, Warenbezeichnungen u.s.w. in diesem Werk berechtigt auch ohne besondere Kennzeichnung nicht zu der Annahme, dass solche Namen im Sinne der Warenzeichen- und Markenschutzgesetzgebung als frei zu betrachten wären und daher von jedermann benutzt werden dürften.

Verlag: Südwestdeutscher Verlag für Hochschulschriften GmbH & Co. KG
Dudweiler Landstr. 99, 66123 Saarbrücken, Deutschland
Telefon +49 681 37 20 271-1, Telefax +49 681 37 20 271-0
Email: info@svh-verlag.de
Zugl.: Jena, FSU, Diss. 2010

Herstellung in Deutschland:
Schaltungsdienst Lange o.H.G., Berlin
Books on Demand GmbH, Norderstedt
Reha GmbH, Saarbrücken
Amazon Distribution GmbH, Leipzig
ISBN: 978-3-8381-2033-1

Imprint (only for USA, GB)
Bibliographic information published by the Deutsche Nationalbibliothek: The Deutsche Nationalbibliothek lists this publication in the Deutsche Nationalbibliografie; detailed bibliographic data are available in the Internet at http://dnb.d-nb.de.

Any brand names and product names mentioned in this book are subject to trademark, brand or patent protection and are trademarks or registered trademarks of their respective holders. The use of brand names, product names, common names, trade names, product descriptions etc. even without a particular marking in this works is in no way to be construed to mean that such names may be regarded as unrestricted in respect of trademark and brand protection legislation and could thus be used by anyone.

Publisher: Südwestdeutscher Verlag für Hochschulschriften GmbH & Co. KG
Dudweiler Landstr. 99, 66123 Saarbrücken, Germany
Phone +49 681 37 20 271-1, Fax +49 681 37 20 271-0
Email: info@svh-verlag.de

Printed in the U.S.A.
Printed in the U.K. by (see last page)
ISBN: 978-3-8381-2033-1

Copyright © 2010 by the author and Südwestdeutscher Verlag für Hochschulschriften GmbH & Co. KG and licensors
All rights reserved. Saarbrücken 2010

Contents

List of abbreviations 3

1. Introduction 5
   1.1. Transcriptional regulation in procaryotes 5
      1.1.1. Transcription initiation 5
      1.1.2. Regulation by promoter selectivity 6
      1.1.3. Regulation by transcription factors 7
         1.1.3.1. Activation 7
         1.1.3.2. Repression 8
      1.1.4. Regulation of transcription factors 10
   1.2. Catabolite repression in *B. subtilis* 10
      1.2.1. Elements of the CCR-Systems of *B. subtilis* 11
      1.2.2. CcpA-dependant catabolite repression 12
      1.2.3. CcpA-independent catabolite repression 14
   1.3. The transcription factor CcpN 15
   1.4. Purpose of this book 16
   1.5. References 17

2. Transcriptional repressor CcpN from *Bacillus subtilis* compensates asymmetric contact distribution by cooperative binding 24
   2.1. Summary 25
   2.2. Introduction 25
   2.3. Results 26
   2.4. Discussion 36
   2.5. Materials and Methods 41
   2.6. References 45

3. Identification of ligands affecting the activity of the transcriptional repressor CcpN from *Bacillus subtilis* 48
   3.1. Summary 49
   3.2. Introduction 49
   3.3. Results 50
   3.4. Discussion 59
   3.5. Materials and Methods 64
   3.6. References 68

4. The transcriptional repressor CcpN from *Bacillus subtilis* uses different repression mechanisms at different promoters 73
   4.1. Summary 74

  4.2. Introduction 74
  4.3. Results 75
  4.4. Discussion 81
  4.5. Experimental Procedures 83
  4.6. References 85

5. **Search for additional targets of the transcriptional regulator CcpN from *B. subtilis* 88**
  5.1. Summary 89
  5.2. Introduction 89
  5.3. Results and Discussion 90
  5.4. Materials and Methods 98
  5.5. References 100

6. **Summary 103**

List of abbreviations

Acetyl-CoA	Acetyl coenzyme A
ADP	Adenosine diphosphate
ATP	Adenosine triphosphate
B. subtilis	*Bacillus subtilis*
bp	basepair
cAMP	cyclic Adenosine monophosphate
CBS	Cystathione β synthetase
CcpA	catabolite control protein A
CcpN	catabolite control protein of gluconeogenic genes
CCR	carbon catabolite repression
CD	circular dichroism
cre	catabolite responsive element
DNA	deoxyribonucleic acid
DNase I	Deoxyribonuclease I
E. coli	*Escherichia coli*
GapB	Glyceraldehyd-3-phosphate dehydrogenase B
His	Histidine
HPr	histidine-containing protein
K_D	dissociation constant
mRNA	messenger RNA
nt	nucleotide
PckA	Phosphoenolpyruvate carboxykinase
PTS	Phosphotransferase system
RNA	ribonucleic acid
RNAP	RNA polymerase
Ser	Serine
SR1	Small RNA 1
ThyB	Thymidylatsynthase B
e. G.	for example
α-CTD	C-terminal domain of the α-subunit
α-NTD	N-terminal domain of the α-subunit

1. Introduction

1.1. Transcriptional regulation in procaryotes

1.1.1. Transcription initiation

During their life, bacteria face a variety of different environmental conditions, to which they must react accordingly. While smaller differences in e.g. osmolarity or intracellular metabolite concentration can be regulated by the action of the corresponding porines or metabolic enzymes, larger and continuing changes require a more intense regulation. Changes in the environmental conditions usually call for changes in the protein composition of the cell, leaving bacteria with only a limited number of possible regulatory mechanisms. First, the amount of protein can be regulated, either by control of synthesis or control of degradation, or second, the amount of the corresponding mRNA can be adjusted. This can also be achieved by control of synthesis or degradation, with the latter representing the more economic variant, since the synthesis of mRNA already represents an energetic effort to the cell.

Regulation of RNA synthesis usually occurs during transcription initiation, or in more rare cases, during elongation. Transcription initiation is a complex process, consisting of a sequence of specific steps (Record *et al.*, 1996; Figure 1). During the first step, the bacterial RNA polymerase (RNAP), an enzyme consisting of several subunits, binds to a promoter region. The polymerase itself consists of a β and a β' subunit, which form the catalytic centre (Korzheva *et al.*, 2000). Assembly of these subunits is supports by the N-terminal domain of two α subunits (Blatter *et al.*, 1994), while their C-terminal domains have supporting function in promoter recognition and binding (Gourse *et al.*, 2000). The σ factor is ultimately responsible for the recognition of the promoter sequence and recruitment of the RNAP to the promoter (Wösten, 1998). Beside these required ones, several subunits can be associated with the RNAP, supporting its function but are not imperative for it. Examples are the ω subunit, which mediates correct folding of the β subunit as a chaperone (Hampsey, 2001), or the δ subunit from *Bacillus subtilis*, which increases transcription efficiency and plays a not yet fully elucidated role during Sporulation (Gao & Aronson, 2004). The promoter region consists of four sequence elements: A -10 region and a -35 region, which can be found at every promoter and are recognised by domains 2 and 4 of the σ factor (Busby & Ebright, 1994; Murakami *et al.*, 2002). For these regions, the consensus sequences TTGACA (-35) and TATAAT (-10) have been determined. Furthermore, an extended -10 region, a TG_n motif 3-4 bp in length (Sanderson *et al.*, 2003) as well as so-called UP-elements, AT rich regions upstream of the -35 region, can be present (Ross *et al.*, 2001). By binding of RNAP to the promoter, the closed complex is formed (RPc), which is then transformed into a complex, in which RNAP is bound more closely to the DNA (RPc2). Subsequently, catalysed by the domains 2 and 4 of the σ factor, the DNA is melted in the region from -10 to +4, creating the open complex (RPo) (Tsujikawa *et al.*, 2002). Upon incorporation of the first nucleoside triphosphates, this complex is transformed into the initiation complex, which creates small, 3-10 nt long abortive transcripts which result from failed attempts of RNAP to exit the promoter. When promoter clearance finally succeeds, RNAP forms the elongation complex. This final step is irreversible, while all other preceding steps are complete

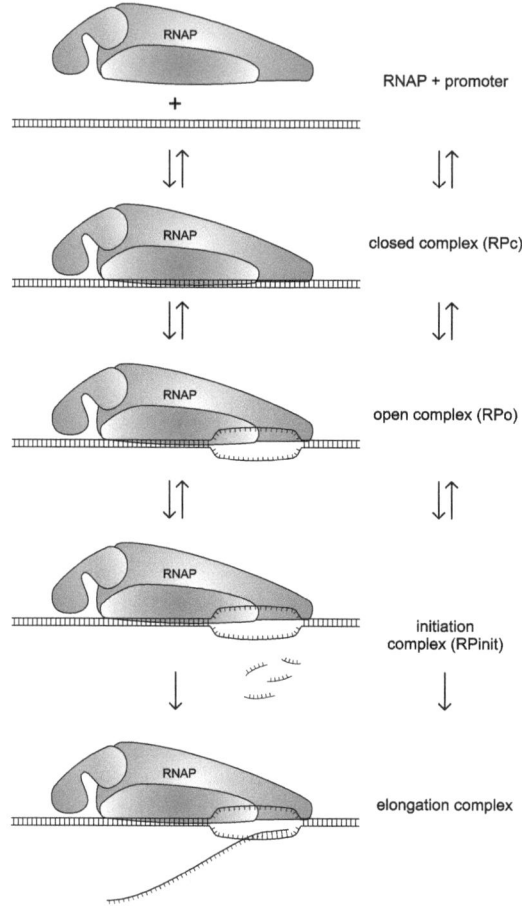

reversible. At each promoter, every step of transcription initiation is characterised by a specific activation energy, making the step with the highest activation energy the rate limiting step.

The regulation of transcription initiation can not only occur at each of these steps, but also in every direction: proteins that reduce the activation energy of a certain step act as transcriptional activators and those that increase the activation energy act as transcription repressors.

Figure 1: Phases of transcription initiation

Schematic picture of the different complexes formed during transcription initiation. All steps except the transition into the elongation complex are reversible.

1.1.2. Regulation by promoter selectivity

Regulation of transcription can be performed at a basal level through the promoter sequence. Promoters whose sequence is close to the consensus sequence, are usually more efficient than those with diverging sequences. However, since this type of regulation is anchored within the DNA sequence, it can only be used to partition the RNAP to different promoters and regulate general transcript levels, but not for individual regulation.

A special position among transcriptional regulators is occupied by σ factors. Different organisms possess different amount of σ factors, ranging from one single factor in *Mycoplasma genitalium* to as much as 63 different factors in *Streptomyces coelicolor* (Gruber & Gross, 2003). Each σ factor has its own requirements regarding the sequence of the -10 and -35 region (sequences mentioned above are for the main σ factor), allowing the RNAP to be guided to specific sets of promoters by certain σ factors. Regulation of σ factor activity can be achieved by its synthesis or degradation, or by so-called anti σ factors. These proteins bind specific σ factors and subsequently prevent σ factor binding by the RNAP (Hughes & Mathee, 1998).

1.1.3. Regulation by transcription factors

1.1.3.1. Activation

Activation of transcription initiation is mainly performed during formation of the closed complex (Figure 2). Promoters whose sequence is greatly divergent from the consensus, are predestined for this type of regulation, since they possess a weak innate promoter activity. Generally, activators can be divided into three classes: Class I activators recruit RNAP to a promoter by interacting with the α-CTD. Such activators bind upstream of the -35 region, but do not require fixed operator positions, since the α-CTD is connected to the RNAP by a flexible linker. Examples for such activators are the cAMP receptor protein CRP at the *E. coli lac* promoter (Ebright, 1993) or protein p4 from the *B. subtilis* phage Φ29 at the A3 promoter (Nuez *et al.*, 1992; Mencía *et al.*, 1996). Class II activators are proteins that interact with domain 4 of the σ factor (Dove *et al.*, 2003). Since the σ factor is not flexible regarding positioning at the promoter, such activators have operator sequences centred around -41.5. One example for such an activator is the

(a) Class I activation

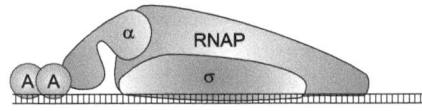

(b) Class II activation

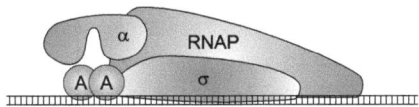

(c) Class III activation

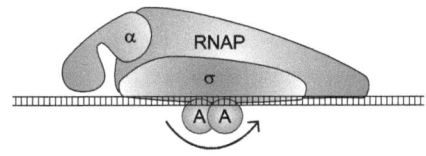

Figure 2: Mechanisms of transcription activation

Activators are marked with an "A" and shown as dimers, since they mostly act as dimers. RNAP denotes the RNA polymerase, α the α subunit.

(a) Class I activation. The activator binds upstream of the promoter and recruits RNAP through contacts with the α-CTDs.

(b) Class II activation. The activator binds at position -41.5 and interacts with domain 4 of the σ factor.

(c) Activation by conformational change. The activator binds between the -35 and the -10 region and rearranges them, allowing RNAP to bind.

CI protein of the bacteriophage λ at the λ PRM promoter (Nickels *et al.*, 2002). In some cases, class II activators have been reported to contact the α-NTD instead of the σ factor (Busby & Ebright, 1997). Class III activators can be found at promoters, at which the -10 and -35 region show a suboptimal orientation, e.g. because the spacer region is too long or too short. Upon binding, the activator can change the spatial structure of the DNA and bring the relevant promoter elements into a more favourable spatial orientation. Members of the MerR family of activators, e.g. BmrR, use this type of activation (Heldwein *et al.*, 2001; Brown *et al.*, 2003). Alternatively, activators that bind far upstream of the RNAP and interact with it by forming a DNA loop are sometimes regarded as class III activators. The formation of the DNA loop can be enhanced by yet another activator.

Promoters that are recognised by σ^{54} from *E. coli* form a special group. These promoters are usually limited in formation of open complexes and require activators that specifically stabilise open complexes (Buck *et al.*, 2000).

1.1.3.2. Repression

In contrast to activators, which almost exclusively enhance formation of the closed complex, repressors can utilise every step during transcription initiation (Figure 3). The simplest possibility of repression is the steric hindrance of RNAP binding, which is mostly achieved by operator sequences that overlap the -10 or -35 region. Examples for this case are the Lac repressor bound to the O1 operator at the *lac* promoter (Schlax *et al.*, 1995) or the p4 protein from Φ29 that has already been introduced as an activator, which can also repress RNAP binding at the A2b promoter (Rojo & Salas, 1991). Beside direct steric interference, RNAP binding can also be prevented by other mechanisms. The GalR repressor for example can bind at different operators that do not overlap with the promoter region and subsequently form a DNA loop mediated by repressor interactions that prevents RNAP binding (Choy & Adhya, 1996). Another possibility is a function as anti-activator, which can be found in the case of the CytR-regulated *deo* promoter in *E. coli*. Here, the CytR repressor binds with the help of two CRP dimers, which are bound as activators and therewith prevents the interaction between CRP and the α-CTDs of RNAP, leaving RNAP unable to be recruited to the promoter (Valentin-Hansen *et al.*, 1996; Shin *et al.*, 2001).

At many promoters, the progression from the closed to the open complex is influenced by repressors. Such repressors can bind to operators that overlap the promoter region, yet allow the simultaneous binding of RNAP. An example for this case is the MerR repressor at the *merT* promoter, for which an inhibition of open complex formation was shown *in vitro* and *in vivo* (Heltzel *et al.*, 1990), or the Spo0A protein from *B. subtilis* at the *abrB* promoter (Greene & Spiegelman, 1996). In addition to this, some repressors of open complex formation, like KorB at the *korABF* promoter (Williams *et al.*, 1993), have operator sites that do not overlap the promoter sequence. Special cases are σ^{54} controlled promoters, which, as already mentioned, require activators to allow open complex formation. At these promoters, repressors can bind between the activator and the RNAP, and, upon induction of DNA bending, prevent interaction between these two proteins. The Nac regulator from *Klebsiella aerogenes* (Feng *et al.*, 1995) or the global regulator CcpA from *B. subtilis* at the *lev* operon should be mentioned as examples (Martin-Vestraete *et al.*, 1995).

(a) Inhibition of promoter binding

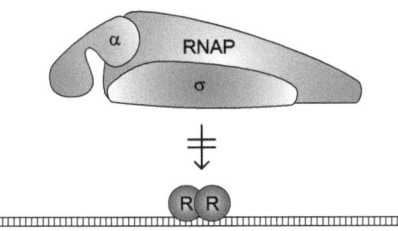

(b) Inhibition of open complex formation

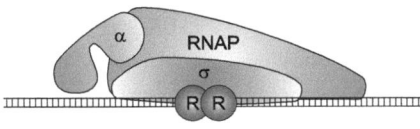

(c) Inhibition of transcription initiation

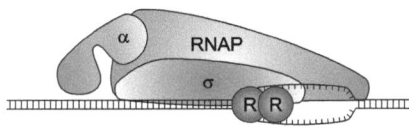

(d) Inhibition of elongation complex formation

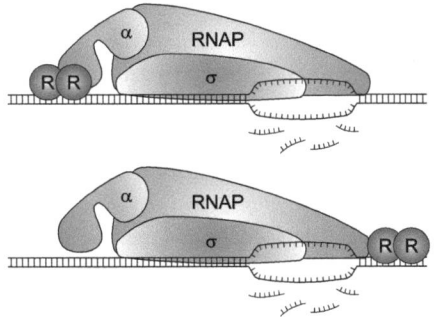

Figure 3: Mechanisms of transcription repression

Activators are marked with an "R" and shown as dimers, since they mostly act as dimers. RNAP denotes the RNA polymerase, α the α subunit and σ the σ factor.

(a) Inhibition of RNAP binding. The repressor blocks the binding site for RNAP by binding in the promoter region.

(b) Inhibition of open complex formation. Though possessing overlapping binding sites, RNAP and the repressor can bind simultaneously. However, the presence of the repressor does not allow formation of open complexes

(c) Inhibition of transcription initiation. RNAP and the repressor can bind simultaneously and a stable open complex is formed, but the repressor prevents transcription. Neither abortive nor full transcripts are formed.

(d) Inhibition of open complex formation. The repressor arrests the RNAP at the promoter, either by direct interaction or by forming a transcriptional roadblock. Abortive transcripts can be formed, but the RNAP cannot leave the promoter.

For the inhibition of the following step, the formation of the initiation complex and the synthesis of abortive transcripts, only few cases have been reported in literature so far. Such a mechanism has been shown for the H-NS protein at the *rrnB* P1 promoter (Schröder & Wagner, 2000) or for the FIS protein at the *gyrB* promoter (Schneider *et al.*, 1999). In both cases the formation of open complexes has been shown, but no transcripts were detectable.

A lot more common is an inhibition of the transition to the elongation complex. In these cases, abortive transcripts are readily formed, but the RNAP is not able to leave the promoter. Such a phenomenon can also be observed at promoters, where all four relevant sequence elements match or closely resemble the consensus sequence. This leads to a very efficient binding of RNAP to the promoter, however, RNAP is unable to leave the promoter afterwards (Ellinger *et al.*, 1994). Because of this, normal promoters often represent a compromise between efficient RNAP recruitment and efficient promoter clearance. Repressors which usually bind highly specific and also tightly to DNA, can mimic the above mentioned effect by interacting with RNAP and causing its arrest at the promoter. Only few examples exist for this mechanism, like the protein p4 of phage Φ29, which has already been introduced as activator and repressor (Monsalve *et al.*, 1996). It is able, in addition to the aforementioned effects, to interact with RNAP and stabilise it at the A2c promoter, preventing transition into an elongation complex. The GalR repressor also works after the principle (Choy *et al.*, 1995). Another possibility to prevent the formation of an elongation complex is the binding of a repressor downstream of +1. This repressor does not need to interact with the RNAP, but represents a roadblock through its tight interaction with DNA, which cannot be overcome by the RNAP. This type of regulation can be found at the *treP* gene in *B. subtilis*, which is negatively regulated by CcpA (Ujiie *et al.*, 2009). Furthermore, an artificial construct, where the *lac* operator was placed downstream of a promoter which is not normally regulated by LacR, showed that this type of repression can theoretically by mediated by all sufficiently tight binding repressors (Lopez *et al.*, 1998).

1.1.4. Regulation of transcription factors

Since transcription factors must influence transcription only under certain conditions, their activity itself also needs to be regulated. This can happen by regulation of synthesis and degradation. However, more economic possibilities exist, which can also take effect in a much shorter time scale. Lots of transcription factors are parts of bacterial two component systems, consisting of a sensor kinase and a response regulator. The kinase detects signals from the environment or the cytoplasm and phosphorylates the regulator, which only then can bind to DNA, or, in other cases, is already bound to DNA but gets active only after phosphorylation, like DeoR from *B. subtilis* (Zeng & Saxild, 1999). Another possibility is the binding of small effector molecules, which can reach from ions, like Ni^{2+} in the case of NikR from *E. coli* (Fauquant *et al.*, 2006), to different metabolites. Like before, the metabolite might induce DNA binding of the regulator either by presence or absence, or an already bound regulator can be transferred into an active conformation.

1.2. Catabolite repression in *B. subtilis*

The metabolism of most bacteria is suited to exploit a variety of different nutrient sources. However, for the utilisation of these sources, highly specialised enzymes are required that are not used in basic metabolism. Prokaryotes have established mechanisms to ensure that they can on the one hand use a large diversity of substances, but on the other hand primarily use those substances whose degradation can be achieved with minimal effort. For most organisms, this substance is glucose, since it can easily be degraded via glycolysis. The mechanism that ensures the preferred use of glucose while shutting down alternative catabolic pathways is called carbon catabolite repression (CCR). In *E. coli*, the most important Gram negative model organism, cAMP, which is enriched in the cell during glucose starvation, has been identified as the signal molecule for this mechanism (Perlman *et al.*, 1969). It can be bound by the cAMP receptor protein (CRP), which in turn can activate alternative metabolic pathways as an activator. Most of these pathways are also controlled by a repressor, which allows the expression of a specific pathway only in the presence of the corresponding nutrient source, to ensure specificity in their regulation. An example for this is the LacI repressor at the *lac* operon, which is removed from the DNA in the presence of lactose (Lewis, 2005).

1.2.1. Elements of the CCR system of *B. subtilis*

In *B. subtilis*, the most important Gram positive model organism, neither cAMP could be detected during glucose starvation, or a homologue to CRP, excluding the *E. coli* model of CCR. In contrast, CCR in *B. subtilis* is mediated by a *cis* active element, the *cre* sequence (catabolite responsive element) (Nicholson *et al.*, 1987; Weickert *et al.*, 1990), and a *trans* active factor, CcpA (catabolite control protein A) (Henkin *et al.*, 1991). CcpA usually acts as a repressor, and its specific binding to *cre* has also been demonstrated *in vitro* (Fujita *et al.*, 1995). An overview over the catabolite repression system can be found in figure 4. In total, over 50 *cre* sequences have been identified in *B. subtilis* and their participation in the regulation of genes and operons has been shown experimentally. Bioinformatics analysis showed the existence of 150 putative *cres*, who together regulate about 300 genes. Out of the 50 experimentally verified *cres*, the consensus sequence WTGAAARCGYTTWNN has been derived (Miwa & Fujita, 1990), whereat it has been discovered that most of the naturally occurring *cres* show small deviations from the consensus, which is necessary for their function (Miwa *et al.*, 2000). Since the transcription factor binding the *cre* sequence, CcpA, is constitutively expressed (Miwa *et al.*, 1994), it became clear that at least one more factor is involved in CCR of *B. subtilis*.

This factor has been identified as the HPr protein (histidine-containing protein) some time later (Deutscher *et al.*, 1994). HPr fulfils several complex tasks in the metabolism of *B. subtilis*. The protein itself has 2 distinct phosphorylation sites, one at histidine 15 and one at serine 46. Both phosphorylation states are mutually exclusive. Position His15 is important for the transport of sugars from the surrounding medium into the cell, because from here, the phosphate residue can be transferred to a sugar transporter from the phosphoenol pyruvate-dependant sugar phosphotransferase system (PTS), and from there to the corresponding sugar. The sugars transported by this system are called PTS sugars and constitute the main nutrient source of *B. subtilis* (Postma *et al.*, 1993; Reizer, 1996). Whenever these sugars are not present, HPr phosphorylated at His15 can then phosphorylate other enzymes like glycerol kinase and antiterminators of alternative metabolic pathways, which eventually leads to their activation (Tortosa & Le Coq, 1995; Schnetz *et al.*, 1996; Darbon *et al.*, 2002).

1.2.2. CcpA-dependant catabolite repression

Phosphorylation of HPr at Ser46 has a completely different function. A phosphorylation at this site is catalysed by HPr kinase/phosphatase, which itself is activated by high concentrations of fructose-1,6-bisphosphate, being an indicator of good nutrient conditions (Reizer *et al.*, 1998). With this phosphorylation state (HPr-Ser46-P), HPr can interact with the transcription factor CcpA and is able to increase its *cre* element binding capability dramatically (Deutscher *et al.*, 1995; Jones *et al.*, 1997). Phosphorylation at Ser46 is reversible and can be removed by the phosphatase activity of HPr kinase/phosphatase. This activity is stimulated by the presence of phosphate ions, which indicate a low intracellular ATP level and poor metabolic conditions (Hanson *et al.*, 2002). This explains how the constitutively expressed CcpA is able to act depending on the metabolic status of the cell. The CcpA-HPr-Ser46-P complex can now regulate transcription at all promoters controlled by a *cre* element. Depending on the position of the *cre* element, different modes of repression or activation can be observed. *Cre* elements upstream of the promoter region often function as activators and require direct interaction between CcpA and RNAP. Examples for this are the *ackA* gene, coding for acetate kinase (Turinsky *et al.*, 1998), the *pta* gene, coding for phosphotransacetylase (Shin *et al.*, 1999) or the *ilvB* gene, which encodes the large subunit of acetolactate synthase (Tojo *et al.*, 2005). The meaning of these three genes will be explained later in the text.

Cre elements overlapping with the promoter region often work through steric hindrance of RNAP-DNA interaction and mediate catabolite repression. Examples for this comprise the *amyE*, *bglP* and *dctP* genes, coding for degradation enzymes or transporters of alternative nutrient sources like starch, disaccharides or various C_4 dicarboxylates (Nicholson *et al.*, 1987; Krüger *et al.*, 1996; Asai *et al.*, 2000). At *cre*-controlled promoters, one can often observe an inhibition of transcription elongation, occurring when the *cre* element is placed downstream of the transcription start site. This type of repression can be found at the genes for citrate synthase (*citZ*) (Kim *et al.*, 2002), a magnesium dependant citrate transporter (*citM*) (Yamamoto *et al.*, 2000) or a transporter for trehalose (*treP*) (Schöck & Dahl, 1996).

During growth in a medium containing sufficient amounts of PTS sugars, the complete metabolism of *B. subtilis* is altered dramatically, as one can see looking at the genes activated and repressed by CcpA. Under normal growth conditions, carbohydrate sources are fed into the appropriate part of glycolysis and eventually completely degraded in the citrate cycle, to ensure a maximal ATP-yield per nutrient molecule. However, when an abundance of PTS sugars like glucose or fructose are present, the organism primarily performs glycolysis, for it might not be the most efficient, but the fastest way of ATP production. This leads to a significant increase in pyruvate and acetyl-CoA levels, since acetyl-CoA cannot enter the citrate cycle due to inhibition of the citrate synthase by CcpA. The cell now has the problem to dispose the accumulating metabolites (Sauer & Eikmanns, 2005). Acetyl-CoA is converted to acetate by the enzymes Pta and AckA, both activated by CcpA. The acetate is then transported out of the cell into the surrounding medium, from where it can be taken up again, if the preferred sugar source is exhausted. Another part of acetyl-CoA is consumed by the fatty acid synthesis, since good nutritional conditions lead to increased growth rates and therewith an increased need for cell membrane parts. A part of the

(a) presence of glucose

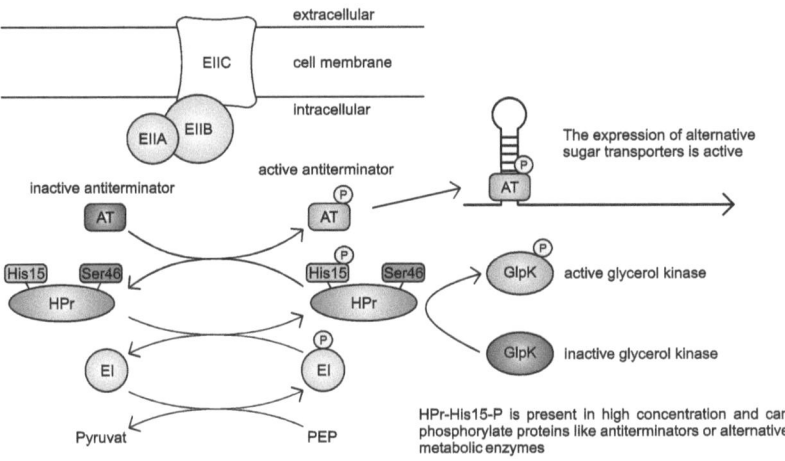

HPr-His15-P is present in high concentration and can phosphorylate proteins like antiterminators or alternative metabolic enzymes

(b) absence of glucose

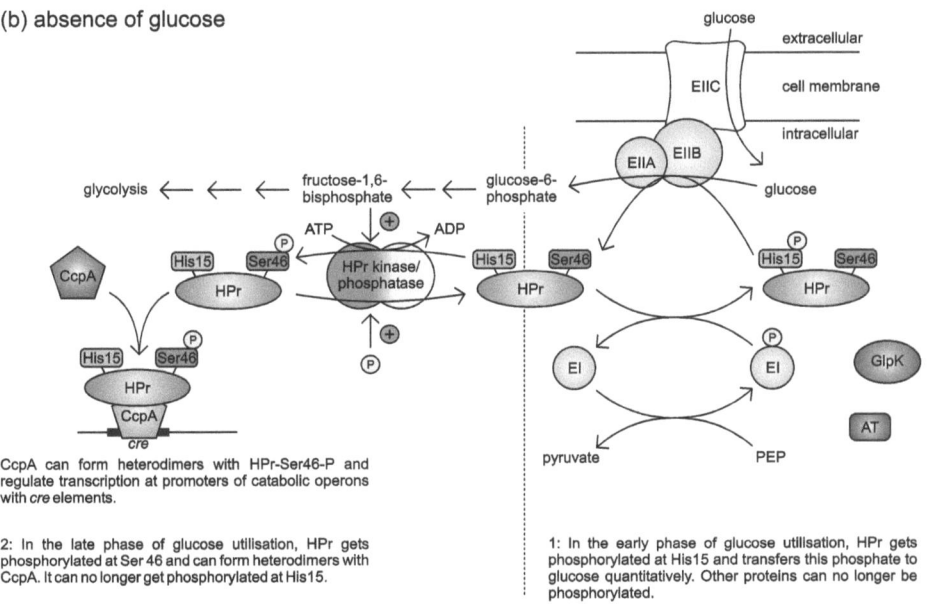

CcpA can form heterodimers with HPr-Ser46-P and regulate transcription at promoters of catabolic operons with *cre* elements.

2: In the late phase of glucose utilisation, HPr gets phosphorylated at Ser 46 and can form heterodimers with CcpA. It can no longer get phosphorylated at His15.

1: In the early phase of glucose utilisation, HPr gets phosphorylated at His15 and transfers this phosphate to glucose quantitatively. Other proteins can no longer be phosphorylated.

Figure 4: Simplified overview over the catabolite repression system in *B. subtilis*

The figure legend can be found on the next page.

Figure 4: Simplified overview over the catabolite repression system in *B. subtilis*

(a) Regulation in the absence of glucose. Due to the absence of a preferred sugar, the concentration of HPr-His15-P is high and the phosphate can be transferred to regulators of alternative catabolic operons, like antiterminators or catabolic enzymes, which in turn become active. EI and EIIA/B/C: Enzymes of the phosphotransferase system.

(b) Regulation in the presence of glucose. The phosphate from HPr-His15-P is transferred to glucose quantitatively, preventing the phosphorylation of other proteins. Alternative metabolic pathways are shut down.

(1) Upon accumulation of glycolysis intermediates like FPB, the kinase activity of HPr kinase/phosphatase is stimulated and HPr phosphorylated at position Ser46. HPr can now no longer phosphorylate glucose, but can form heterodimers with CcpA and regulate the transcription at promoters with cre elements.

(2) When ATP levels in the cell fall and phosphate concentration rises, the phosphatase activity of HPr kinase/phosphatase is activated. HPr can now no longer interact with CcpA, and cre elements can no longer be bound. The phosphorylation states Ser46-P and His15-P are mutually exclusive.

accumulating pyruvate is converted to acetoine and transported into the medium by the products of the *alsSD* operon, which is also activated by CcpA (Renna *et al.*, 1993). The other part is converted to acetolactate by the acetolactate synthase and then used in the synthesis of branched-chain amino acids to satisfy the protein needs of rapidly growing cells (Shivers & Sonenshein, 2005).

Bioinformatics analysis of the *B. Subtilis* genome showed a protein with a sequence similar to HPr: Crh (Galinier *et al.*, 1997). It has been demonstrated that this protein can be phosphorylatd by the HPr kinase/phosphatase and that it can interact with CcpA. The exact function of this protein is still unknown, but there is evidence that it takes over the role of HPr during growth on non-carbohydrate substrates like succinate or glutamate (Warner & Lolkema 2003, Görke *et al.*, 2004).

1.2.3. CcpA independent catabolite repression

Although CcpA controls most of the genes in *B. Subtilis* that are regulated by the CCR system, microarray analyses showed that several genes are repressed by glucose in the medium in a *ccpA* knockout strain or in a strain where HPr cannot be phosphorylated at Ser46 (Yoshida *et al.*, 2001; Lulko *et al.*, 2007). This finding let to the discovery of new catabolite control proteins, which will be introduced in this chapter.

CcpB, a protein paralogue to CcpA, is involved in the catabolite repression of some operons for the utilisation of gluconate and xylose and appears to be active primarily during growth on solid media. However, since its discovery 10 years ago, it has not been investigated further (Chauvaux *et al.*, 1998).

CcpC represses the genes *citZ*, *citB* and *citC*, which encode the first three enzymes of the citrate cycle (Jourlin-Castelli *et al.*, 2000). The *ccpC* gene is negatively controlled by CcpC and CcpA, reducing the repressor concentration to a low but sufficient level for efficient repression

(Kim *et al.*, 2002; Kim *et al.*, 2003). Citrate is a negative regulator of CcpC, ensuring that the citrate cycle is activated once significant amounts of citrate accumulate in the cell. This and the omission of repression by CcpA lead to increased levels of CcpC, assuring that the repression can become active when citrate gets scarce.

CggR (central glycolytic genes repressor) represses the expression of the *cggR-gapA-pgk-tpi-pgm-eno* operon, encoding the repressor itself as well as all enzymes that are required for the further degradation of C_3 intermediates in glycolysis (Fillinger *et al.*, 2000). Since all of these enzymes, except GapA, are also required for gluconeogenesis, the operon has a second promoter ensuring constant levels of these enzymes within the cell (Ludwig *et al.*, 2001). GapA, together with GapB, present a peculiarity of *B. Subtilis*, since both enzymes catalyse the same reaction, but are only active during glycolysis or gluconeogenesis, respectively. In most other microorganisms, this reaction is catalysed by a single enzyme. The affinity of CggR to its operator is negatively regulated by fructose-1,6-bisphosphate (FBP) (Doan & Aymerich, 2003). Since the concentration of FBP is significantly higher during glycolysis than during gluconeogenesis, it is an eligible signal for the regulation of CggR. As mentioned earlier, FBP also acts as an activator of HPr kinase/phosphatase and thus represents the central signalling molecule of CCR in *B. subtilis*.

1.3. The transcription factor CcpN

The latest discovered transcription factor that acts in catabolite repression is CcpN (catabolite control protein of gluconeogenic genes) and was identified while investigating the regulation of *pckA* (PEP carboxykinase) and *gapB* (glyceraldeyde-3-phosphate dehydrogenase B), both active exclusively during gluconeogenesis (Servant *et al.*, 2005). At the same time, CcpN was discovered independently as the transcriptional regulator of a small regulatory RNA, SR1 (Licht *et al.*, 2005). SR1 is a negative regulator of AhrC, which in turn is a positive regulator of arginine degradation genes and a negative regulator of arginine synthesis genes (Heidrich *et al.*, 2006). It has been shown that SR1 can interact with the *ahrC* mRNA by basepairing, thereby altering their structure and inhibiting translation initiation (Heidrich *et al.*, 2007). For all three genes, *pckA*, *gapB* and *sr1*, a strong repression in the presence of any sugar source that is fed into glycolysis, like glucose, fructose or glycerol, has been demonstrated. Servant *et al.* showed that CcpN is essential for efficient growth under glycolytic conditions and that *ccpN* knockout strains are limited in their growth rate. Under gluconeogenic conditions, a weak positive effect on growth rate was detectable (Servant *et al.*, 2005).

The *ccpN* gene is transcribed together with the *yqfL* gene as a bicistronic mRNA, which is constitutively expressed. Homologues of *ccpN* have been found in many firmicutes, like *B. halodurans*, *Geobacillus stearothermophilus*, *B. cereus* and *B. anthracis*. The function of the YqfL protein has not yet been elucidated, but is has been shown that it acts as a positive regulator of *ccpN-yqfL* operon expression. However, it does not respond to the metabolic state of the cell. The specific binding of CcpN to its operators has been shown by EMSA and DNase I footprinting. These experiments showed that CcpN apparently has two operators at the *gapB* and *sr1* promoter, and one extended operator site at the *pckA* promoter. In addition, the induction of several DNase I hypersensitive sites upon CcpN binding has been observed, indicating a change in DNA structure

(Servant *et al.*, 2005; Licht *et al.*, 2005). Licht *et al.*, using one operator from the *sr1* promoter, showed that a single operator site is sufficient for CcpN binding. With the help of EMSAs with mutated oligonucleotides, the following provisional consensus sequence for CcpN operators was derived: DDDTGTGYYATACTRDK. A search for this sequence in the *B. subtilis* genome revealed a variety of genes with a CcpN operator in their promoter region, among them *gapB* and *pckA*. However, only those two beside the *sr1* operator showed CcpN binding activity.

Meanwhile, the biophysical characteristics of CcpN have also been studied in detail. It has been shown that CcpN exists as a dimer in the cell (Zorrilla *et al.*, 2008). However, the gathered information is in part contradictory to previous results. As an example, the binding stoichiometry for CcpN has been determined to be two molecules of CcpN at the *gapB* operator, and four at the *pckA* operator. This is in contrast to the observations from EMSA experiments, which showed identical protein DNA complex sizes at all operators. This obvious contradiction has, however, not been discussed by the authors.

In contrast, the regulation of the intracellular metabolic fluxes controlled by CcpN, is understood much better (Tännler *et al.*, 2008). By measuring the intracellular concentration of different metabolites in wild type and *ccpN* knockout strains it has been shown that in a *ccpN* knockout strain, extensive amounts of energy are consumed in a "futile cycle" due to the uncontrolled expression of *pckA*. Here, PEP is converted to pyruvate by pyruvate kinase, which in turn is metabolised to oxaloacetate using ATP. The oxaloactetate is then converted back to PEP by PckA under ATP usage. This cycle proceeds uncontrolled, consuming ATP and draining oxaloactetate from the citrate cycle. In fact, the lack of oxaloacetate is the main cause for the growth defect of a *ccpN* knockout strain, because the deficiency in oxaloacetate inhibits aspartate synthesis. When additional aspartate is provided in the growth medium during glycolytic conditions, the growth rate is converted back to normal. Beside futile cycling and oxaloactetate deprivation, a *ccpN* knockout strain is characterised by a dramatically increased metabolic flux through the pentose phosphate pathway. The overexpression of *gapB* is responsible for this effect, since it converts the majority of synthesised 1,3-bisphospho glycerate immediately back to glyceraldehyde-3-phosphate. This causes a metabolic jamming in the upper part of glycolysis, which is solved by the cell by rerouting significant amounts of FBP into the pentose phosphate pathway.

1.4. Purpose of this book

Bacteria have to adapt to constantly changing environmental conditions, indicated by changes in nutrient availability as well as temperature and osmolarity variations. The adaption to different nutrient sources is accomplished by a process called catabolite repression. This process has – despite 20 years of research – not yet been fully understood in *B. subtilis*. The discovery of new factors, like CcpN or the Crh protein, constantly helps in accomplishing the big picture. In this book, the latest discovered factor, CcpN, will be characterised in detail to gain new insights into its DNA binding characteristics and its mode of action. Furthermore, it is shown how one can gather in-depth information on transcription factors by sophisticated, yet equipment-wise undemanding laboratory techniques.

1.5. References

1. Asai, K., Baik, S. H., Kasahara, Y., Moriya, S. & Ogasawara, N. (2000) Regulation of the transport system for C4-dicarboxylic acids in *Bacillus subtilis*. *Microbiology* **146**: 263-271.
2. Blatter, E. E., Ross, W., Tang, H., Gourse, R. L. & Ebright, R. H. (1994) Domain organisation of RNA-polymerase α-subunit: C-terminal 85 amino-acids constitute a domain capable of dimerisation and DNA-binding. *Cell* **78**: 889-896.
3. Brown, N. L., Stoyanov, J. V., Kidd, S. P. & Hobman, J. L. (2003) The MerR family of transcriptional regulators. *FEMS Microbiol. Rev.* **27**: 145-163.
4. Buck, M., Gallegos, M.-T., Studholme, D. J., Guo, Y. & Gralla, J. D. (2000) The bacterial enhancer-dependent σ^{54} (σ^{N}) transcription factor. *J. Bacteriol.* **182**: 4129-4136.
5. Busby, S. & Ebright R. H. (1994) Promoter structure, promoter recognition, and transcription activation in prokaryotes. *Cell* **79**: 743-746.
6. Busby, S. & Ebright, R. H. (1997) Transcription activation at Class II CAP-dependent promoters. *Mol. Microbiol.* **23**: 853-859.
7. Chauvaux, S., Paulsen, I. T. & Saier, M. H. Jr. (1998) CcpB, a novel transcription factor implicated in catabolite repression in *Bacillus subtilis*. *J. Bacteriol.* **180**: 491-497.
8. Choy, H. & Adhya. S. (1996) Negative control. In *Escherichia coli and Salmonella typhimurium: Cellular and Molecular Biology*. Edited by Neidhardt, F. C., Ingraham, J. L., Low, K. B., Magasanik, B., Schaechter, M. & Umbarger, H. E. Washington DC: American Society for Microbiology; 1287-1299.
9. Choy, H. E., Park, S. W., Aki, T., Parrack, P., Fujita, N., Ishihama, A. & Adhya, S. (1995) Repression and activation of transcription by Gal and Lac repressors: involvement of alpha subunit of RNA polymerase. *EMBO J.* **14**: 4523-4529.
10. Darbon, E., Servant, P., Poncet, S. & Deutscher, J. (2002) Antitermination by GlpP, catabolite repression via CcpA and inducer exclusion triggered by P-GlpK dephosphorylation control *Bacillus subtilis glpFK* expression. *Mol. Microbiol.* **43**: 1039-1052.
11. Deutscher, J., Küster, E., Bergstedt, U., Charrier, V. & Hillen, W. (1995) Protein kinase-dependent HPr/CcpA interaction links glycolytic activity to carbon catabolite repression in gram-positive bacteria. *Mol. Microbiol.* **15**: 1049-1053.
12. Deutscher, J., Reizer, J., Fischer, C., Galinier, A., Saier, M. H. Jr. & Steinmetz, M. (1994) Loss of protein kinase-catalyzed phosphorylation of HPr, a phosphocarrier protein of the phosphotransferase system, by mutation of the *ptsH* gene confers catabolite repression resistance to several catabolic genes of *Bacillus subtilis*. *J. Bacteriol.* **176**: 3336-3344.
13. Doan, T. & Aymerich, S. (2003) Regulation of the central glycolytic genes in *Bacillus subtilis*: binding of the repressor CggR to its single DNA target sequence is modulated by fructose-1,6-bisphosphate. *Mol. Microbiol.* **47**: 1709-1721.

14. Dove, S. L., Darst, S. A. & Hochschild, A. (2003) Region 4 of σ as a target for transcription regulation. *Mol. Microbiol.* **48**: 863-874.
15. Ebright, R. H. (1993) Transcription activation at Class I CAP-dependent promoters. *Mol. Microbiol.* **8**: 797-802.
16. Ellinger, T., Behnke, D., Bujard, H. & Gralla, J. D. (1994) Stalling of *Escherichia coli* RNA polymerase in the +6 to +12 region *in vivo* is associated with tight binding to consensus promoter elements. *J. Mol. Biol.* **239**: 455-465.
17. Fauquant, C., Diederix, R. E., Rodrigue, A., Dian, C., Kapp, U., Terradot, L., Mandrand-Berthelot, M. A. & Michaud-Soret, I. (2006) pH dependent Ni(II) binding and aggregation of *Escherichia coli* and *Helicobacter pylori* NikR. *Biochimie* **88**: 1693-1705.
18. Feng, J., Gos, T. J., Bender, R. A. & Ninfa, A. J. (1995) Repression of the *Klebsiella aerogenes nac* promoter. *J. Bacteriol.* **177**: 5535-5538.
19. Fillinger, S., Boschi-Muller, S., Azza, S., Dervyn, E., Branlant, G. & Aymerich, S. (2000) Two glyceraldehyde-3-phosphate dehydrogenases with opposite physiological roles in a nonphotosynthetic bacterium. *J. Biol.Chem.* **275**: 14031-14037.
20. Fujita, Y., Miwa, Y., Galinier, A. & Deutscher, J. (1995) Specific recognition of the *Bacillus subtilis gnt* cis-acting catabolite-responsive element by a protein complex formed between CcpA and seryl-phosphorylated HPr. *Mol. Microbiol.* **17**: 953-960.
21. Galinier, A., Haiech, J., Kilhoffer, M. C., Jaquinod, M., Stülke, J., Deutscher, J. & Martin-Verstraete, I. (1997) The *Bacillus subtilis crh* gene encodes a HPr-like protein involved in carbon catabolite repression. *Proc. Natl. Acad. Sci. USA* **94**: 8439-8444.
22. Gao, H. & Aronson, A. I. (2004) The delta subunit of RNA polymerase functions in sporulation. *Curr. Microbiol.* **48**: 401-404.
23. Görke, B., Fraysse, L. & Galinier, A. (2004) Drastic differences in Crh and HPr synthesis levels reflect their different impacts on catabolite repression in *Bacillus subtilis*. *J. Bacteriol.* **186**: 2992-2995.
24. Gourse, R. L., Ross, W. & Gaal T. (2000) UPs and downs in bacterial transcription initiation: the role of the α subunit of RNA polymerase in promoter recognition. *Mol. Microbiol.* **37**: 687-695.
25. Greene, E. A. & Spiegelman, G. B. (1996) The Spo0A protein of *Bacillus subtilis* inhibits transcription of the *abrB* gene without preventing binding of the polymerase to the promoter. *J. Biol. Chem.* **271**: 11455-11461.
26. Gruber, T. M. & Gross, C. A. (2003) Multiple sigma subunits and the partitioning of bacterial transcription space. *Annu. Rev. Microbiol.* **57**: 441-466.
27. Hampsey, M. (2001) Omega meets its match. *Trends Genet.* **17**: 190-191.

28. Hanson, K. G., Steinhauer, K., Reizer, J., Hillen, W. & Stülke, J. (2002) HPr kinase/phosphatase of *Bacillus subtilis*: expression of the gene and effects of mutations on enzyme activity, growth and carbon catabolite repression. *Microbiology* **148**: 1805-1811.

29. Heidrich, N., Chinali, A., Gerth, U. & Brantl, S. (2006) The small untranslated RNA SR1 from the *Bacillus subtilis* genome is involved in the regulation of arginine catabolism. *Mol. Microbiol.* **62**: 520-536.

30. Heidrich, N., Moll, I. & Brantl, S. (2007) *In vitro* analysis of the interaction between the small RNA SR1 and its primary target *ahrC* mRNA. *Nucleic Acids Res.* **35**: 4331-4346.

31. Heldwein, E. E. & Brennan, R. G. (2001) Crystal structure of the transcription activator BmrR bound to DNA and a drug. *Nature* **409**: 378-382.

32. Heltzel, A., Lee, I. W., Totis, P. A. & Summers, A. O. (1990) Activator-dependent preinduction binding of σ^{70}-RNA polymerase at the metal-regulated *mer* promoter. *Biochemistry* **29**: 9572-9584.

33. Henkin, T. M., Grundy, F. J., Nicholson, W. L. & Chambliss, G. H. (1991) Catabolite repression of alpha-amylase gene expression in *Bacillus subtilis* involves a trans-acting gene product homologous to the *Escherichia coli* lacI and galR repressors. *Mol. Microbiol.* **5**: 575-584.

34. Hughes, K. & Mathee, K. (1998) The anti-sigma factors. *Annu. Rev. Microbiol.* **52**: 231-286.

35. Jones, B. E., Dossonnet, V., Küster, E., Hillen, W., Deutscher, J. & Klevit, R. E. (1997) Binding of the catabolite repressor protein CcpA to its DNA target is regulated by phosphorylation of its corepressor HPr. *J. Biol. Chem.* **272**: 26530-26535.

36. Jourlin-Castelli, C., Mani, N., Nakano, M. M. & Sonenshein, A. L. (2000) CcpC, a novel regulator of the LysR family required for glucose repression of the *citB* gene in *Bacillus subtilis*. *J. Mol. Biol.* **295**: 865-878.

37. Kim, H. J., Jourlin-Castelli, C., Kim, S. I. & Sonenshein, A. L. (2002) Regulation of the *Bacillus subtilis ccpC* gene by CcpA and CcpC. *Mol. Microbiol.* **43**: 399-410.

38. Kim, H. J., Roux, A. & Sonenshein, A. L. (2002) Direct and indirect roles of CcpA in regulation of *Bacillus subtilis* Krebs cycle genes. *Mol. Microbiol.* **45**: 179-190.

39. Kim, S. I., Jourlin-Castelli, C., Wellington, S. R. & Sonenshein, A. L. (2003) Mechanism of repression by *Bacillus subtilis* CcpC, a LysR family regulator. *J. Mol. Biol.* **334**: 609-624.

40. Korzheva, N., Mustaev, A., Kozlov, M., Malhotra, A., Nikiforov, V., Goldfarb, A. & Darst, S. A. (2000) A structural model for transcription elongation. *Science* **289**: 619-625

41. Krüger, S., Gertz, S. & Hecker, M. (1996) Transcriptional analysis of *bglPH* expression in *Bacillus subtilis*: evidence for two distinct pathways mediating carbon catabolite repression. *J. Bacteriol.* **178**: 2637-2644.

42. Lewis, M. (2005) The Lac repressor. *Crit. Rev. Biol.* **328**: 521-548.

43. Licht, A., Preis, S. & Brantl, S. (2005) Implication of CcpN in the regulation of a novel untranslated RNA (SR1) in *Bacillus subtilis*. *Mol. Microbiol.* **58**: 189-206.
44. Lopez, P. J., Guillerez, J., Sousa, R. & Dreyfus, M. (1998) On the mechanism of inhibition of phage T7 RNA polymerase by *lac* repressor. *J. Mol. Biol.* **276**: 861-875.
45. Ludwig, H., Homuth, G., Schmalisch, M., Dyka, F. M., Hecker, M. & Stülke, J. (2001) Transcription of glycolytic genes and operons in *Bacillus subtilis*: evidence for the presence of multiple levels of control of the *gapA* operon. *Mol. Microbiol.* **41**: 409-422.
46. Lulko, A. T., Buist, G., Kok, J., Kuipers & O. P. (2007) Transcriptome analysis of temporal regulation of carbon metabolism by CcpA in *Bacillus subtilis* reveals additional target genes. *J. Mol. Microbiol. Biotechnol.* **12**: 82-95.
47. Martin-Vestraete, I., Stülke, J., Klier, A. & Rapoport, G. (1995) Two different mechanisms mediate catabolite repression of the *Bacillus subtilis* levanase operon. *J. Bacteriol.* **177**: 6919-6927.
48. Mencía, M., Monsalve, M., Rojo, F. & Salas, M. (1996) Transcription activation by phage Φ29 protein p4 is mediated by interaction with the α subunit of *Bacillus subtilis* RNA polymerase. *Proc. Natl. Acad. Sci. USA* **93**: 6616-6620.
49. Miwa, Y. & Fujita, Y. (1990) Determination of the cis sequence involved in catabolite repression of the *Bacillus subtilis gnt* operon; implication of a consensus sequence in catabolite repression in the genus Bacillus. *Nucleic Acids Res.* **18**: 7049-7053.
50. Miwa, Y., Nakata, A., Ogiwara, A., Yamamoto, M. & Fujita, Y. (2000) Evaluation and characterization of catabolite-responsive elements (*cre*) of *Bacillus subtilis*. *Nucleic Acids Res.* **28**: 1206-1210.
51. Miwa, Y., Saikawa, M. & Fujita Y. (1994) Possible function and some properties of the CcpA protein of *Bacillus subtilis*. *Microbiology* **140**: 2567-2575.
52. Monsalve, M., Mencía, M., Salas, M. & Rojo, F. (1996) Protein p4 represses phage Φ29 A2c promoter by interacting with the α subunit of *Bacillus subtilis* RNA polymerase. *Proc. Natl. Acad. Sci. USA* **93**: 8913-8918.
53. Murakami, K. S., Masuda, S., Campbell, E. A., Muzzin, O. & Darst, S. A. (2002) Structural basis of transcription initiation: an RNA polymerase holoenzyme–DNA complex. *Science* **296**, 1285-1290.
54. Nicholson, W. L., Park, Y. K., Henkin, T. M., Won, M., Weickert, M. J., Gaskell, J.A. & Chambliss, G. H. (1987) Catabolite repression-resistant mutations of the *Bacillus subtilis* alpha-amylase promoter affect transcription levels and are in an operator-like sequence. *J. Mol. Biol.* **198**: 609-618.
55. Nickels, B. E., Dove, S. L., Murakami, K. S., Darst, S. A. & Hochschild, A. (2002) Protein–protein and protein–DNA interactions of σ^{70} region 4 involved in transcription activation by λcI. *J. Mol. Biol.* **324**: 17-34.

56. Nuez, B., Rojo, F. & Salas, M. (1992) Phage Φ29 regulatory protein p4 stabilizes the binding of the RNA polymerase to the late promoter in a process involving direct protein-protein contacts. *Proc. Natl. Acad. Sci. USA* **89**: 11401-11405.

57. Perlman, R. L., De Crombrugghe, B. & Pastan, I. (1969) Cyclic AMP regulates catabolite and transient repression in *E. coli. Nature* **223**: 810-812

58. Postma, P. W., Lengeler, J. W. & Jacobson, G. R. (1993) Phosphoenolpyruvate: carbohydrate phosphotransferase systems of bacteria. *Microbiol. Rev.* **57**: 543-594.

59. Record, M. T. Jr., Reznikoff, W. S., Craig, M. L., McQuade, K. L. & Schlax, P. J. (1996) *Escherichia coli* RNA polymerase ($E\sigma^{70}$), promoters, and the kinetics of the steps of transcription initiation. In *Escherichia coli and Salmonella typhimurium: Cellular and Molecular Biology.* Edited by Neidhardt, F. C., Ingraham, J. L., Low, K. B., Magasanik, B., Schaechter, M. & Umbarger, H. E. Washington DC: American Society for Microbiology; 792-820.

60. Reizer, J., Bergstedt, U., Galinier, A., Küster, E., Saier, M. H. Jr., Hillen, W., Steinmetz, M. & Deutscher, J. (1996) Catabolite repression resistance of *gnt* operon expression in *Bacillus subtilis* conferred by mutation of His-15, the site of phosphoenolpyruvate-dependent phosphorylation of the phosphocarrier protein HPr. *J. Bacteriol.* **178**: 5480-5486.

61. Reizer, J., Hoischen, C., Titgemeyer, F., Rivolta, C., Rabus, R., Stülke, J., Karamata, D., Saier, M. H. Jr. & Hillen, W. (1998) A novel protein kinase that controls carbon catabolite repression in bacteria. *Mol. Microbiol.* **27**: 1157-1169.

62. Renna, M. C., Najimudin, N., Winik, L. R. & Zahler, S. A. (1993) Regulation of the *Bacillus subtilis alsS, alsD,* and *alsR* genes involved in post-exponential-phase production of acetoin. *J. Bacteriol.* **175**: 3863-3875.

63. Rojo, F. & Salas, M. (1991) A DNA curvature can substitute phage Φ29 regulatory protein p4 when acting as a transcriptional repressor. *EMBO J.* **10**: 3429-3438.

64. Ross, W., Ernst, A. & Gourse, R. L. (2001) Fine structure of *E. coli* RNA polymerase-promoter interactions: α subunit binding to the UP element minor groove. *Genes Dev.* **15**: 491-506.

65. Sanderson, A., Mitchell, J. E., Minchin, S. D. & Busby, S. J. (2003) Substitutions in the *Escherichia coli* RNA polymerase σ^{70} factor that affect recognition of extended −10 elements at promoters. *FEBS Lett.* **544**: 199-205.

66. Sauer, U. & Eikmanns, B. J. (2005) The PEP-pyruvate-oxaloacetate node as the switch point for carbon flux distribution in bacteria. *FEMS Microbiol. Rev.* **29**: 765-794.

67. Schlax, P. J., Capp, M. W. & Record, M.T. Jr. (1995) Inhibition of transcription initiation by lac repressor. *J. Mol. Biol.* **245**: 331-350.

68. Schneider, R., Travers, A., Kutateladze, T. & Muskhelishvili, G. (1999) A DNA architectural protein couples cellular physiology and DNA topology in *Escherichia coli*. *Mol. Microbiol.* **34**: 953-964.

69. Schnetz, K., Stülke, J., Gertz, S., Krüger, S., Krieg, M., Hecker, M. & Rak, B. (1996) LicT, a *Bacillus subtilis* transcriptional antiterminator protein of the BglG family. *J. Bacteriol.* **178**: 1971-1979.

70. Schöck, F. & Dahl, M. K. (1996) Analysis of DNA flanking the *treA* gene of *Bacillus subtilis* reveals genes encoding a putative specific enzyme IITre and a potential regulator of the trehalose operon. *Gene* **175**: 59-63.

71. Schröder, O. & Wagner, R. (2000) The bacterial DNA-binding protein H-NS represses ribosomal RNA transcription by trapping RNA polymerase in the initiation complex. *J. Mol. Biol.* **298**: 737-748.

72. Servant, P., Le Coq, D. & Aymerich, S. (2005) CcpN (YqzB), a novel regulator for CcpA-independent catabolite repression of *Bacillus subtilis* gluconeogenic genes. *Mol. Microbiol.* **55**: 1435-1451.

73. Shin, B. S., Choi, S. K. & Park, S. H. (1999) Regulation of the *Bacillus subtilis* phosphotransacetylase gene. *J. Biochem.* **126**: 333-339.

74. Shin, M., Kang, S., Hyun, S. J., Fujita, N., Ishihama, A., Valentin-Hansen, P. & Choy, H. E. (2001) Repression of deoP2 in *Escherichia coli* by CytR: conversion of a transcription activator into a repressor. *EMBO J.* **20**: 5392-5399.

75. Shivers, R. P. & Sonenshein, A. L. (2005) *Bacillus subtilis ilvB* operon: an intersection of global regulons. *Mol. Microbiol.* **56**: 1549-1559.

76. Tännler, S., Fischer, E., Le Coq, D., Doan, T., Jamet, E., Sauer, U. & Aymerich, S. (2008) CcpN controls central carbon fluxes in *Bacillus subtilis*. *J. Bacteriol.* **190**: 6178-6187

77. Tojo, S., Satomura, T., Morisaki, K., Deutscher, J., Hirooka, K. & Fujita, Y. (2005) Elaborate transcription regulation of the *Bacillus subtilis ilv-leu* operon involved in the biosynthesis of branched-chain amino acids through global regulators of CcpA, CodY and TnrA. *Mol. Microbiol.* **56**: 1560-1573.

78. Tortosa, P. & Le Coq, D. (1995) A ribonucleic antiterminator sequence (RAT) and a distant palindrome are both involved in sucrose induction of the B*acillus subtilis sacXY* regulatory operon. *Microbiology* **141**: 2921-7292.

79. Tsujikawa, L., Tsodikov, O. V. & deHaseth, P. L. (2002) Interaction of RNA polymerase with forked DNA: evidence for two kinetically significant intermediates on the pathway to the final complex. *Proc. Natl. Acad. Sci. USA* **99**: 3493-3498.

80. Turinsky, A. J., Grundy, F. J., Kim, J. H., Chambliss, G. H. & Henkin, T. M. (1998) Transcriptional activation of the *Bacillus subtilis ackA* gene requires sequences upstream of the promoter. *J. Bacteriol.* **180**: 5961-5967.

81. Ujiie, H., Matsutani, T., Tomatsu, H., Fujihara, A., Ushida, C., Miwa, Y., Fujita, Y., Himeno, H. & Muto, A. (2009) Trans-translation is involved in the CcpA-dependent tagging and degradation of TreP in *Bacillus subtilis*. *J. Biochem.* **145**: 59-66.

82. Valentin-Hansen, P., Søgaard-Andersen, L. & Pedersen, H. (1996) A flexible partnership: the CytR anti-activator and the cAMP-CRP activator protein, comrades in transcription control. *Mol. Microbiol.* **20**: 461-466.

83. Warner, J. B. & Lolkema, J. S. (2003) A Crh-specific function in carbon catabolite repression in *Bacillus subtilis*. *FEMS Microbiol. Lett.* **220**: 277-280.

84. Weickert, M. J. & Chambliss, G. H. (1990) Site-directed mutagenesis of a catabolite repression operator sequence in *Bacillus subtilis*. *Proc. Natl. Acad. Sci. USA* **87**: 6238-6242.

85. Williams, D. R., Motallebi-Veshareh, M. & Thomas, C. M. (1993) Multifunctional repressor KorB can block transcription by preventing isomerization of RNA polymerase-promoter complexes. *Nucleic. Acids. Res.* **21**: 1141-1148.

86. Wösten, M. M. (1998) Eubacterial sigma-factors. *FEMS Microbiol. Rev.* **22**: 127–150.

87. Yamamoto, H., Murata, M. & Sekiguchi, J. (2000) The CitST two-component system regulates the expression of the Mg-citrate transporter in *Bacillus subtilis*. *Mol. Microbiol.* **37**: 898-912.

88. Yoshida, K., Kobayashi, K., Miwa, Y., Kang, C. M., Matsunaga, M., Yamaguchi, H., Tojo, S., Yamamoto, M., Nishi, R., Ogasawara, N., Nakayama, T. & Fujita, Y. (2001) Combined transcriptome and proteome analysis as a powerful approach to study genes under glucose repression in *Bacillus subtilis*. *Nucleic Acids Res.* **29**: 683-692.

89. Zeng, X. & Saxild, H. H. (1999) Identification and characterization of a DeoR-specific operator sequence essential for induction of *dra-nupC-pdp* operon expression in *Bacillus subtilis*. *J. Bacteriol.* **181**: 1719-1727.

90. Zorrilla, S., Ortega, A., Chaix, D., Alfonso, C., Rivas, G., Aymerich, S., Lillo, M. P., Declerck, N. & Royer, C. A. (2008) Characterization of the control catabolite protein of gluconeogenic genes repressor by fluorescence cross-correlation spectroscopy and other biophysical approaches. *Biophys. J.* **95**: 4403-4415.

2. Transcriptional repressor CcpN from *Bacillus subtilis* compensates asymmetric contact distribution by cooperative binding

Andreas Licht* & Sabine Brantl

AG Bakteriengenetik, Friedrich-Schiller-Universität Jena, D-07743 Germany

Published in: *Journal of Molecular Biology*, **364**: 434-448 (2006)

*corresponding author

2.1. Summary

Carbon catabolite repression in *B. subtilis* is carried out mainly by the major regulator CcpA. In contrast, sugar dependent repression of three genes, *sr1* encoding a small nontranslated RNA, and two genes coding for gluconeogenic enzymes, *gapB* and *pckA*, is mediated by the recently identified transcriptional repressor CcpN. Since previous DNase I footprinting yielded only basic information on the operator sequences of CcpN, chemical interference footprinting studies were performed for a precise contact mapping. Methylation interference, potassium permanganate and hydroxylamine footprinting were used to identify all contacted residues in both strands in the three operator sequences. Furthermore, ethylation interference experiments were performed to identify phosphate residues essential for CcpN binding. Here, we show that each operator has two binding sites for CcpN, one of which was always contacted stronger than the other. The three sites that exhibited close contacts were very similar in sequence, with only a few slight variations, whereas the other three corresponding sites showed several deviations. Gel retardation assays with purified CcpN demonstrated that the differences in contact number and strength correlated well with significantly different K_D values for the corresponding single binding sites. However, quantitative DNase I footprinting of whole operator sequences revealed cooperative binding of CcpN that, apparently, compensated the asymmetric contact distribution. Based on these data, possible consequences for the repression mechanism of CcpN are discussed.

2.2. Introduction

Although many bacteria – among them *B. subtilis* – are able to utilise a vast number of other nutrients,[1,2] glucose is their preferred carbon source.[3] Therefore, cells need to shut-down other catabolic pathways in the presence of glucose to maximise the energy yield.[4] This is accomplished by the so-called catabolite repression. In *Escherichia coli*, catabolite repression is mediated by the central signalling molecule cAMP and its receptor protein CRP.[5,6] By contrast, *B. subtilis* does neither encode a CRP homologue nor does it produce detectable amounts of cAMP under aerobic conditions.[7] Instead, catabolite repression in *B. subtilis* is mainly carried out by the concerted action of CcpA and HPr-Ser46-P, which can interact to form a transcriptional repressor or activator, regulating genes involved in carbon catabolism.[8] However, it has recently been shown that at least two genes, *gapB* and *pckA*, are downregulated in the presence of glucose, independent of CcpA.[9,10] Instead, they are regulated by a novel transcriptional repressor found by transposon mutagenesis screening for derepression of *gapB* and, therefore, named CcpN (control catabolite protein of gluconeogenic genes).[11] The *gapB* gene encodes the rare isotype B of glyceraldehyde-3-phosphate dehydrogenase, and its gene product catalyses the conversion of 1,3-bisphosphoglycerate to glyceraldehyde-3-phosphate, but only during gluconeogenesis.[9] The *pckA* gene codes for another enzyme required for the synthesis of glucose from Krebs cycle intermediates, PEP carboxykinase, which catalyses the conversion of oxaloacetate to phosphoenolpyruvate.[12]

The *ccpN* gene is cotranscribed with the *yqfL* gene resulting in a bicistronic mRNA. It was shown that this operon is not autoregulated, but constitutively expressed under both glycolytic and gluconeogenic conditions.[11] Homologues of CcpN have been found in the genomes of other Bacilli, e.g. *B. halodurans, B. cereus, B. anthracis* and *Geobacillus stearothermophilus*, and in different Firmicutes.[11]

Recently, a third gene regulated by CcpN, *sr1*, has been discovered. This gene codes for a small untranslated RNA, SR1, which has been identified by a systematic search for small RNAs within intergenic regions of the *B. subtilis* genome.[13] *Sr1* was expressed during gluconeogenesis, but repressed under glycolytic conditions. The trans-acting factor responsible for sugar mediated repression was identified as CcpN.[13] Previous DNase I footprinting experiments for all three known CcpN operators indicated different locations of the binding regions relative to the transcription start site.

The aim of the present work was to investigate the interaction between CcpN and its operator regions in more detail using chemical interference footprinting. These experiments showed that contact strength varied greatly depending on the sequence of a given site. Gel retardation assays with single binding sites confirmed these observations. However, quantitative DNase I footprinting experiments with DNA fragments of all three genes spanning the corresponding complete operator sequences indicated cooperative binding of CcpN. The possible impact of these results on the repression mechanism is discussed.

2.3. Results

Chemical interference footprinting experiments were performed with CcpN-His$_5$ (containing five additional C-terminal histidine residues) purified from an *E. coli* overexpression strain. EMSAs (electrophoretic mobility shift assays) have verified that His-tagged CcpN shows the same binding properties as wild-type CcpN and northern blots showed that it can exert the function of wild-type CcpN in a *ccpN* knockout strain (data not shown). All nucleotide numbers in the following paragraphs refer to the transcription start sites. The coding strand is always termed 'top strand', the noncoding strand 'bottom strand'.

Methylation interference

To determine guanosines and adenosines contacted by CcpN, methylation interference experiments were performed. DNA fragments were modified at purine residues by DMS prior to CcpN binding. Adenine is methylated at position *N3* in the minor groove and guanine at position *N7* in the major groove. Figure 1 shows the positions of the methyl-groups interfering with CcpN binding. The top strand of the *sr1* operator exhibited interference at positions G_{-53} and G_{-51} in site I and, to a weaker extend, at G_{-21} and G_{-19} in site II. Adenosine residues with a major contribution to CcpN binding were only found in site I at the top strand (A_{-48} and A_{-46}), whereas in site II only less close contacts were detected. At the bottom strand, methylation of G_{-45} in site I and G_{-17} in site II interfered with CcpN binding. Only less close contacts to adenosine residues have been found in the bottom strand: A_{-52} and A_{-47} in site I and A_{-20} in site II were contacted by CcpN.

Since the contacts to guanosine residues were in all cases closer than those to adenosines, these results indicate that CcpN contacts the DNA mainly via the major groove with some auxiliary

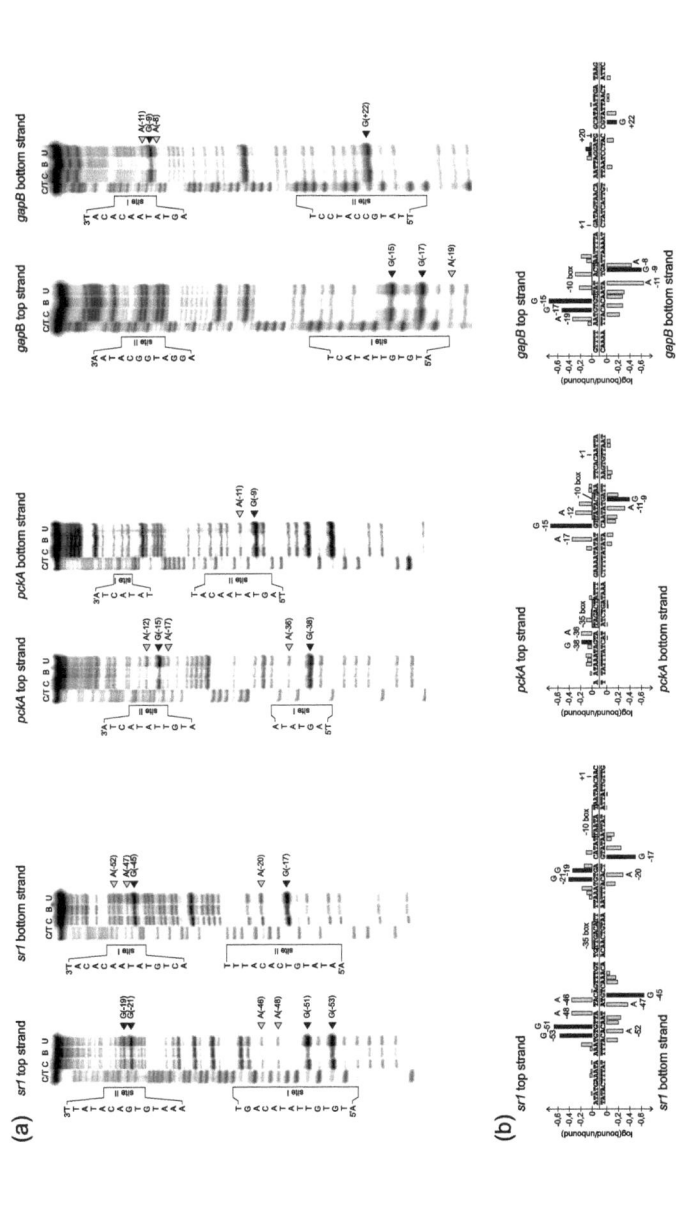

Figure 1: Methylation interference of the *srl*, *pckA*, and *gapB* operators

(a) C/T, Maxam Gilbert C+T sequencing reaction; C, control (protein-free methylated DNA, this lane is equivalent to a Maxam-Gilbert G>A sequencing reaction); B and U, bound and unbound fraction of methylated DNA subjected to binding with CcpN-His$_6$. The numbers in the gels and column diagrams show the positions of the corresponding nucleotides relative to the transcription start site. Binding sites I and II for CcpN have been denoted according to interference footprinting experiments. Close contacts are indicated by black and grey triangles for G and A, respectively.

(b) Column diagrams indicating the relative strength of interference signals for both strands of the three operators. Only positive signals, i.e. signals that indicate contacts, are shown. Measured values were averaged from four independent experiments.

contacts in the minor groove. Furthermore, the contacts in site II were generally less close than those in site I.

In the *pckA* operator, only three contacts to guanosine residues have been observed: G_{-38} in site I and G_{-15} in site II at the top strand as well as G_{-9} in site II at the bottom strand. Binding site II was found to be contacted much stronger than site I. The same was true for contacts to adenosines. Whereas there were some significant contacts in binding site II (A_{-17} and A_{-12} at the top strand and A_{-11} at the bottom strand), only one contacted adenosine was detected in site I (A_{-36} at the top strand). No significant contacts were found in binding site I on the bottom strand.

The *gapB* operator showed a similarly asymmetric contact distribution, but here, contacts were concentrated in binding site I: Close contacts to guanosines (G_{-17} and G_{-15} at the top strand and G_{-9} on the bottom strand) were found, whereas in site II only one less closely contacted guanosine (G_{+22} on the bottom strand) was detected. The same contact distribution was found for adenosine residues. Close and medium contacts were only observed in binding site I (A_{-19} at the top strand and A_{-11}, A_{-8} at the bottom strand). In binding site II, no close contacts to adenosines were observed.

Interestingly, contacts in the *srl* operator were concentrated upstream of the -35 region and, to a lesser extent, in the spacer between -35 and -10, whereas almost no contacts were observed directly within the -35 and -10 regions of p_{SR1}. In contrast, in both the *pckA* and *gapB* operator, close contacts were only found within the -10 region, whereas weak binding sites covered the -35 region and the region downstream from the transcription start site in the *pckA* and *gapB* operator, respectively. Moreover, the interference footprinting revealed that each of the three operators had two CcpN binding sites, although they appeared – due to the short spacer region – as one extended site in the previous DNase I footprints of the *pckA* operator.[11]

Potassium permanganate footprinting

Potassium permanganate footprinting was performed to determine contacts of CcpN to thymidine residues within the three operators. $KMnO_4$, a strong oxidising agent, specifically oxidises thymine, thus impeding protein contacts. In addition to thymine, guanosine is also modified by $KMnO_4$, which results in bands for guanosine residues in the gels. Figure 2 shows the positions of the modified thymidines interfering with CcpN binding. In general, contact distribution correlated well with that found by methylation interference footprinting. The *srl* operator exhibited the following strong interference signals in binding site I: T_{-52} and T_{-50} at the top strand and T_{-48}, T_{-46}, T_{-44} at the bottom strand. However, in contrast to the contacts to guanosines and adenosines, contacts to thymidines (T_{-22} and T_{-20} at the top strand) were slightly closer in binding site II. Significant contacts in binding site II have not been found on the bottom strand.

Both in the *pckA* and *gapB* operators, the positions of contacted thymidines corresponded perfectly to those identified for guanines and adenines by methylation interference, too. The focus of contacts was in site II in the case of *pckA* (five close contacts, see Figure 2, at the top strand and T_{-12} and T_{-10} at the bottom strand), while only less close contacts were found in binding site I at the top strand and no significant contacts at all at the bottom strand. The *gapB* operator exhibited strong interference signals only in site I (mainly T_{-16} and T_{-14} on the top strand and T_{-12} and T_{-10} on the bottom strand), whereas only less close contacts were found in binding site II at the top strand and

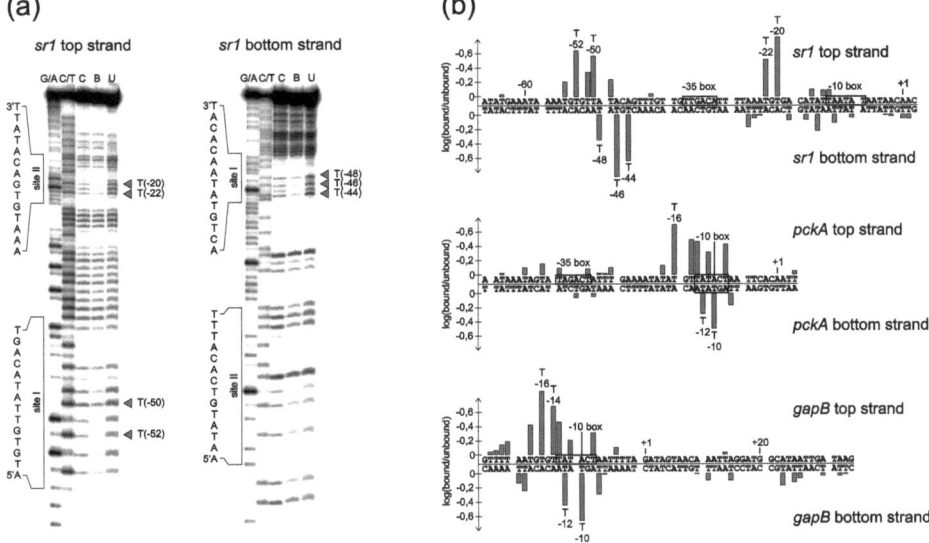

Figure 2: KMnO₄ interference of the *sr1*, *pckA* and *gapB* operators

(a) G/A, Maxam Gilbert G>A sequencing reaction; C/T, Maxam Gilbert C+T sequencing reaction; C, control (protein-free KMnO₄-treated DNA); B and U, bound and unbound fraction of methylated DNA subjected to binding with CcpN-His$_5$. The numbers in the gels and column diagrams show the positions of the corresponding nucleotides relative to the transcription start site. Binding sites I and II for CcpN are designated as in Figure 1. Close contacts are indicated by dark grey triangles. Only the gels for top and bottom strand of the *sr1* operator are shown.

(b) The column diagrams present the relative strength of interference signals for both strands of the three operators. Only positive signals, i.e. signals that indicate contacts, are shown. Measured values are averaged from four independent experiments.

no significant contacts at the bottom strand. In all three operators, thymidines that showed the strongest interference signals were located next to contacted guanosines, together forming the contact center within each binding site.

Hydroxylamine footprinting

NH₂OH footprinting was used to analyse CcpN contacts to cytidines in the three operators. Hydroxylamine, a strong reductive agent, causes ring opening specifically at cytosines and, in this way, interferes with contact formation between protein and DNA. Figure 3 shows that contacts to cytidines were found in all three operators, however, the contact were less close compared to the three other bases. Interference signals of almost equal intensity were found in the *sr1* operator in

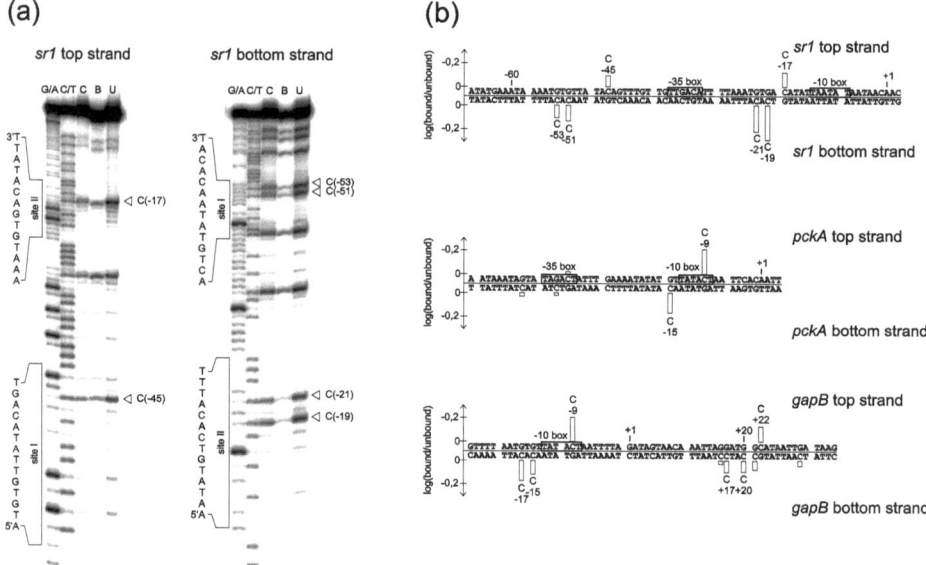

Figure 3: NH₂OH interference of the *sr1*, *pckA* and *gapB* operators

(a) G/A, Maxam Gilbert G>A sequencing reaction; C/T, Maxam Gilbert C+T sequencing reaction; C, control (protein free NH₂OH-treated DNA); B and U, bound and unbound fraction of methylated DNA subjected to binding with CcpN-His$_5$. The numbers in the gels and column diagrams show the positions of the corresponding nucleotides relative to the transcription start site. Binding sites for CcpN are designated as in Figure 1. Close contacts are indicated by white triangles. Only the gels for top and bottom strand of the *sr1* operator are shown.

(b) The column diagrams present the relative strength of interference signals for both strands of the three operators. As above, only positive signals are shown. Measured values are averaged from four independent experiments.

site I (C_{-45} at the top strand and C_{-53} and C_{-51} on the bottom strand) and in site II (C_{-17} at the top strand and C_{-21} and C_{-19} on the bottom strand). The *pckA* operator showed only two contacted cytidines in binding site II and no contacts in binding site I. Interestingly, in the *gapB* operator three contacted cytidines were found in both site I (C_{-9} at the top strand and C_{-15} and C_{-17} at the bottom strand) and site II (C_{+22} at the top strand and C_{+17} and C_{+20} at the bottom strand). However, due to the weak nature of these interference signals, contacts to cytidines do not seem to play an important role in the CcpN-DNA interaction.

Ethylation interference footprinting

To determine phosphate groups of the DNA backbone contacted by CcpN, ethylation interference experiments were carried out. Figure 4 presents the positions at which ethylation interfered with CcpN binding. Both binding sites in the *srl* operator showed only two interference signals: In site I, T_{-50} at the top strand and A_{-47} at the bottom strand were contacted and in site II, A_{-12} and A_{-15} at the bottom strand were contacted. In the *pckA* operator, only in binding site II contacts to the sugar-phosphate backbone were detected. Here, T_{-14} at the top strand and A_{-8} and T_{-10} at the bottom strand exhibited interference signals. The same was found for the *gapB* operator, where only binding site I showed two contacts to T_{-14} at the top strand and T_{-10} at the bottom strand.

Interestingly, the few DNA-backbone contacts were observed in most cases next to a contacted guanosine residue. Obviously, these contacts play only a minor role in the binding of CcpN to its operators. Figure 5 summarises schematically all probed contacts for the three operators.

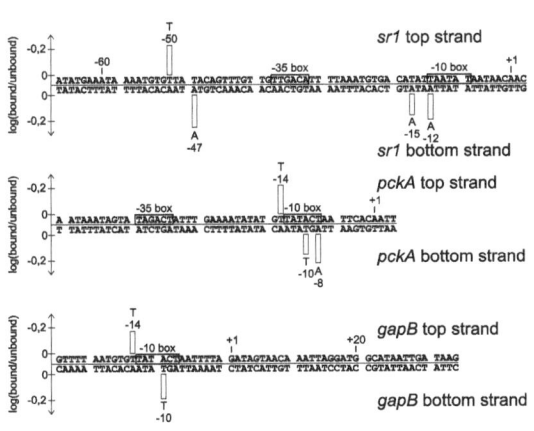

Figure 4: Ethylation interference of the *srl*, *pckA* and *gapB* operators

The column diagrams present the relative strength of interference signals for both strands of the three operators. Only positive signals, i.e. signals that indicate contacts, are shown. Numbers in the column diagrams designate the positions of the corresponding nucleotides relative to the transcription start site. Measured values are averaged from three independent experiments.

EMSA

To determine the apparent equilibrium dissociation constants K_D for the CcpN-DNA complex, 23 bp double-stranded oligonucleotides containing a single CcpN binding site were incubated with increasing concentrations of CcpN-His$_5$ (Figure 6(a)). K_D values were estimated by nonlinear regression using the average data from three independent experiments as described in *Materials and Methods*. The calculated K_D values as well as the binding energy ΔG for the CcpN-DNA interaction for the single sites are summarised in Table 1. Binding energy was calculated with the help of Van't Hoff's reaction isobare $\Delta G = -RT\ln(K)$, where R is the universal gas constant, T the absolute temperature in Kelvin and K the determined equilibrium association constant. The calculated K_D values for the single binding sites corresponded very well to the contacts that were observed by interference footprinting: Binding site I of the *gapB* operator, the one with the most

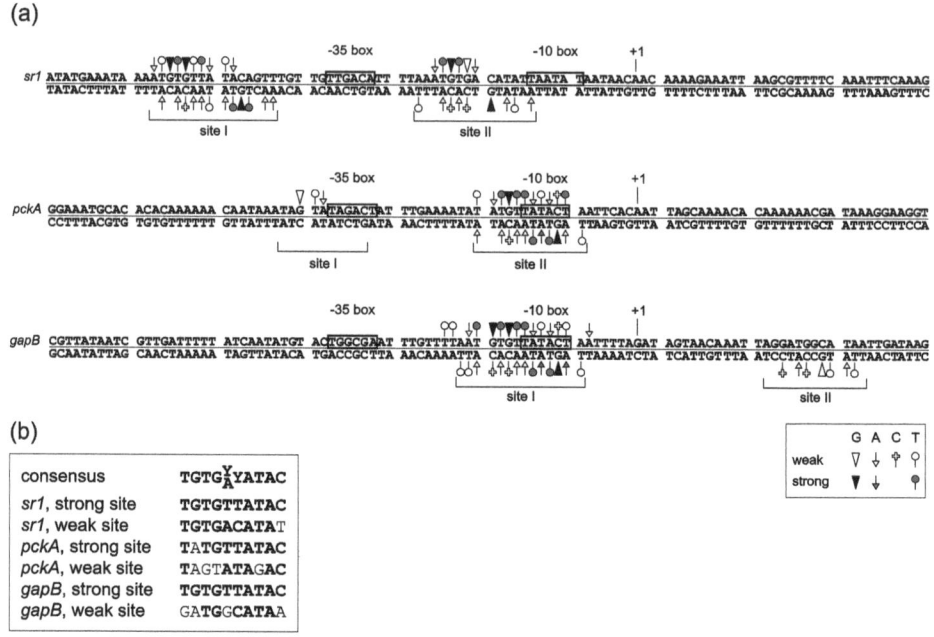

Figure 5: Overview of the contacts in all three operators

(a) Overview of all contacts. Symbols used to indicate contacts to the bases are shown in the box below. Filled symbols denote close contacts (50 % - 100 % compared to the strongest signal), empty symbols represent medium or less close contacts (15 % - 50 % compared to the strongest signal). Contacts with less than 15% relative strength are not shown for better clarity. The -35 and -10 regions are boxed and the transcription start site is indicated. Binding sites I and II are designated based on all interference footprinting experiments.

(b) Alignment of the core sequences of all binding sites. Positions that coincide with the consensus are shown in bold.

and closest contacts (see Figure 1), showed the lowest K_D value indicating a tight protein-DNA interaction, whereas binding site II of *gapB* or site I of *pckA*, both with less close contacts, exhibited high K_D values. Determined dissociation constants ranged from as low as 98 nM (*gapB*, site I) till 4.4 µM (*gapB*, site II).

To test whether the equilibrium dissociation constants differ when using whole operators, double-stranded oligonucleotides containing both CcpN binding sites were incubated with increasing concentrations of CcpN-His$_5$ (Figure 6(b)). K_D values were estimated by nonlinear regression using the average data from three independent experiments as described in *Materials and Methods*. The apparent equilibrium dissociation constants were determined to be 19.3 nM, 15.5 nM and 12.8 nM for the *srl*, *pckA* and *gapB* operator, respectively, and, therewith, correspond well to the values determined by Servant et al.[11] All operators showed significantly lower K_D values than the single sites alone. At the *srl* operator, the average K_D was decreased 30-fold, while the K_D of site I of the *pckA* and site II of the *gapB* operators was decreased 160-fold and 340-fold, respectively.

Quantitative DNase I footprinting

Since in EMSA occupancy of single sites is not detectable, the affinity of CcpN to the single sites within the complete operator was measured by quantitative DNase I footprinting. This technique allows to determine K_D values for site I and site II separately, even if they are located on one DNA fragment. To this end, 89 bp double-stranded oligonucleotides were incubated with increasing concentrations of CcpN-His$_5$, and, after equilibrium was reached, subjected to cleavage with DNase I (Figure 6(c)). In all experiments, the top strand was labelled, since the DNase I cleavage pattern of this strand was more homogeneous than that of the bottom strand. The degree of protection observed corresponded directly to the occupancy of the DNA by CcpN and allowed to calculate the amount of complex formed. Apparent equilibrium dissociation constants were estimated by nonlinear regression using the average data from three independent experiments. The calculated K_D values, the Hill coefficients and the binding energy ΔG for the CcpN-DNA interaction for all single sites are summarised in Table 1. Interestingly, the apparent dissociation constants for the complete operator sequences differed from those found for the investigated single sites and from the results of the footprinting experiments.

Table 1: Apparent dissociation constants and free energies for all CcpN binding sites

	single sites			whole operator		$\Delta\Delta G$	h
binding site	K_D (nM)	ΔG (kJ/mol)	binding site	K_D (nM)	ΔG (kJ/mol)	(kJ/mol)	
srl, site I	420 (±100)	-37.9 (±0.7)	*srl*, site I	80 (±6)	-42.1 (±0.2)	-9.5	1.37
srl, site II	650 (±130)	-36.8 (±0.5)	*srl*, site II	81 (±16)	-42.1 (±0.5)		
pckA, site I	2530 (±310)	-33.2 (±0.3)	*pckA*, site I	145 (±32)	-40.6 (±0.6)	-9.8	1.45
pckA, site II	290 (±60)	-38.8 (±0.5)	*pckA*, site II	115 (±13)	-41.2 (±0.3)		
gapB, site I	98 (±12)	-41.6 (±0.3)	*gapB*, site I	89 (±6)	-41.9 (±0.2)	-9.7	1.49
gapB, site II	4400 (±15)	-31.8 (±0.0)	*gapB*, site II	114 (±48)	-41.2 (±1.0)		

Values were derived from three independent experiments. $\Delta\Delta G$ ($\Delta G_{complete} - \Delta G_{single}$) is the extra free energy that is gained when the two occupied sites are together on one DNA molecule and h is the Hill coefficient.

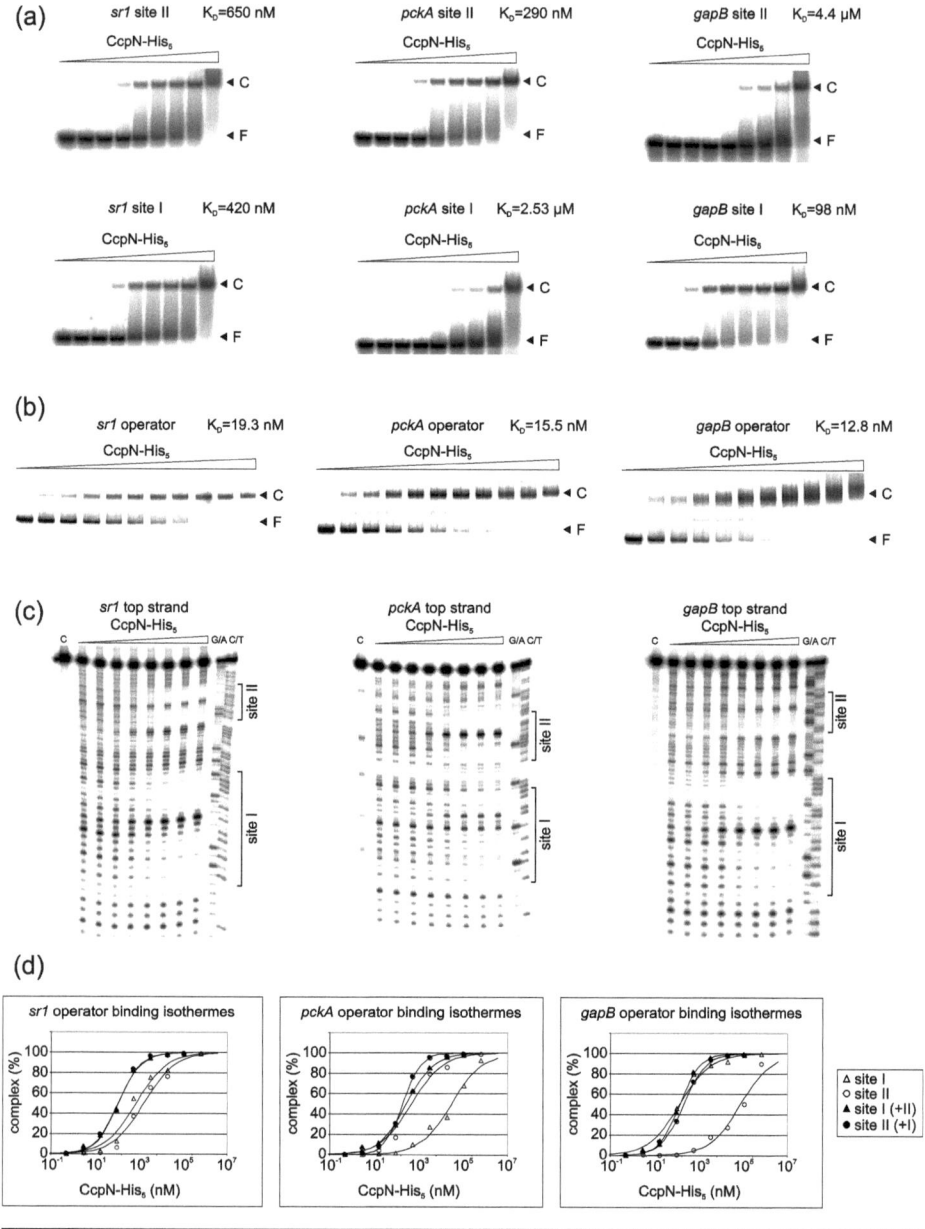

Figure 6: Determination of the binding isothermes for the *sr1*, *pckA* and *gapB* operators

Figure legend on next page.

Figure 6: Determination of the binding isothermes for the *sr1*, *pckA* and *gapB* operators

(a) EMSAs of single CcpN binding sites. 23 bp oligonucleotides were incubated with increasing concentrations of purified CcpN-His$_5$ (CcpN concentration from left to right: 0 nM; 8.1 nM; 27.3 nM; 72.9 nM; 219 nM; 656 nM; 1.97 µM; 5.90 µM; 17.7 µM). F, free DNA; C, CcpN-DNA complex. To allow for a direct comparison with B, EMSAs for binding sites I and II are shown in the same order as the binding sites appear in the DNase I footprinting gels.

(b) EMSAs of whole CcpN operators. 400 bp oligonucleotides were incubated with increasing concentrations of purified CcpN-His$_5$ (CcpN concentration from left to right: 0 nM; 5.2 nM; 7.8 nM; 11.7 nM; 17.6 nM; 26.3 nM; 39.5 nM; 59.3 nM; 88.9 nM; 133 nM; 200 nM). F, free DNA; C, CcpN-DNA complex. The determined K$_D$ values are given in each diagram.

(c) Binding isothermes of the single CcpN binding sites and single-site isothermes for the whole operator sequence. Single sites I and II are represented by empty triangles and circles, respectively. Filled triangles (site I) and circles (site II) designate single-site isothermes of the complete operator. Trend curves shown are averaged from three independent experiments.

(d) Quantitative DNase I footprinting: C, control (uncleaved DNA); G/A, Maxam Gilbert G>A sequencing reaction; C/T, Maxam Gilbert C+T sequencing reaction. 89 bp oligonucleotides containing both CcpN binding sites were incubated with increasing amounts of CcpN-His$_5$ (CcpN concentration from left to right: 0 nM; 8.1 nM; 27.3 nM; 72.9 nM; 219 nM; 656 nM; 1.97 µM; 5.90 µM). Protected regions have been denoted site I and site II.

Cooperativity of CcpN binding

The K$_D$ values for each side in the context of the whole operator were in all cases lower than for the corresponding single sites alone (see Table 1). This was especially true for binding site I of *pckA* and binding site II of the *gapB* operator. The gain in free energy upon CcpN binding to two separated single sites was lower than to two sites in a complete operator, i.e. the occupation of both sites in the operator is cooperative. This was verified by the finding that the K$_D$ values obtained with DNA fragments spanning the whole operator are significantly lower than those obtained with single binding sites. Furthermore, when the values for the single site isothermes in the context of the whole operator were fitted to the Hill equation (see **Materials and Methods**), the shape and the slope of the isothermes changed to a characteristic form for cooperative interactions. Moreover, the Hill coefficient h is in each case >1 (see Table 1), which is a reliable sign for cooperativity. In the case of *sr1*, where the two binding sites have nearly identical K$_D$ values, the affinity of each site was increased by approximately equal amounts. By contrast, when one binding site was much stronger than the other, like in the case of *pckA* and *gapB*, the K$_D$ value for the weaker binding site was improved dramatically (from 4.4 µM to 114 nM for *gapB* site II), but the K$_D$ for the stronger binding site remained mostly unaffected. A comparison of the binding isothermes for the single sites and the single site isothermes for the complete operators that can be found in Figure 6(d) corroborates this conclusion.

Energetic calculations on CcpN-DNA interactions

Quantitative footprinting experiments like those described above were performed at 37 °C and 52 °C. The free energy ΔG was calculated based on three independent experiments. Equation 1 that describes the correlation between free energy (G), enthalpy (H) and entropy (S) can be rearranged to yield equation 2, because ΔH and ΔS are temperature independent. Thereby, T_1 and T_2 are 310.15 K and 325.15 K, respectively, and G_1 and G_2 the free energies at the corresponding temperatures.

$$\Delta G = \Delta H - T\Delta S \quad 1$$

$$\Delta S = \frac{\Delta G_1 - \Delta G_2}{T_2 - T_1} \quad 2$$

Enthalpic and entropic contributions to CcpN-DNA binding were calculated using equation 2 and are summarised in Table 2. The CcpN-DNA interaction shows a small but unfavourable change in entropy which is overcome by a strong enthalpic contribution. This combination of enthalpy and entropy ensures that the CcpN-DNA interaction has nearly the same efficiency at all temperatures that are tolerated by *B. subtilis*.

Table 2: Reaction enthalpy and entropy for the CcpN-DNA interaction

binding site	ΔG (kJ/mol), 37 °C	ΔG (kJ/mol), 52 °C	ΔH (kJ/mol)	ΔS [kJ/(mol*K)]
sr1, site I	-42.3	-41.8	-51.5	-0.03
sr1, site II	-42.7	-41.5	-68.4	-0.08
pckA, site I	-40.4	-39.5	-59.2	-0.06
pckA, site II	-41.7	-41.4	-48.6	-0.02
gapB, site I	-41.7	-41.3	-49.8	-0.03
gapB, site II	-41.2	-38.4	-100.1	-0.18

Quantitative footprinting was performed at 37 °C and 52 °C with DNA fragments carrying the whole operator sequence. The values were derived from three independent experiments.

2.4. Discussion

CcpN binds asymmetrically to its two consecutive binding sites in all three operators

Here we report the high resolution contact probing of the transcriptional repressor CcpN bound to its operator sites. CcpN, which has recently been identified as a repressor active under glycolytic conditions, is known to regulate three genes in *B. subtilis*: *sr1*, encoding a small untranslated RNA[13] and genes for two gluconeogenic enzymes, *pckA* and *gapB*.[11] Using chemical interference footprinting with different chemical probes, we determined the bases contacted by CcpN in all three operators (summarised in Figure 5).

In all cases, two binding sites were identified, one of which was always contacted stronger than the other one. In the following, this site will be referred to as the 'strong site', whereas the other one will be designated 'weak site'. Within all binding sites, core regions can be defined that resemble the consensus binding sequence TGTG(Y/A)YATAC that was previously determined for CcpN.[13] A comparison of all core regions with this consensus is presented in Figure 5(b).

In the *sr1* operator, the upstream binding site (site I, the strong site) was found to be contacted in a slightly stronger manner, but both binding sites showed extensive contacts especially to guanosine and thymidine residues (Figure 5) and less close contacts to adenosines, cytidines and to the sugar-phosphate backbone (Figures 1, 3 and 4). Moreover, both core regions conform well to the consensus. By contrast, contact distribution was found to be completely different in the other two operators. In the case of *pckA*, the majority of contacts were concentrated in the downstream binding site (site II, the strong site), where close contacts to all bases except for cytidines were found (Figure 5). At site I, the weak site, only few and less close contacts were detected. Whereas the core region of the strong site again corresponded well to the consensus sequence, the core of the weak binding site deviated significantly from the consensus. Furthermore, although only one extended site appeared in the *pckA* DNase I footprint,[11] chemical interference revealed that the *pckA* operator consists of two binding sites, too.

Similar results have been found for the *gapB* promoter, except that the upstream site (site I) proved to be the strong site. Like in the case of *pckA*, most and closest contacts were found in the strong site in the consensus-like core region, whereas site II showed only low similarity to the consensus sequence. A series of *gapB* operator mutants tested by Servant *et al.*[11] can be evaluated in the light of the data published here: They found that a $T_{-11} \rightarrow A$ mutation, located in the strong site, severely inhibited CcpN binding, which can be explained by the close contact that we observed to the adenosine residue on the complementary strand. Moreover, this position has shown to be invariant in the previously determined consensus sequence.[13] The was holds true for the $A_{23} \rightarrow G$ mutation, which concerns an invariant base in the weak site. However, the observed effect was not that pronounced, since the contribution of this position is not that great in this case. By contrast, the $T_{-14} \rightarrow G$ mutation showed almost no effect on the CcpN-DNA interaction, despite the close contacts that we mapped for this position. However, this site has been shown to be more variant in the consensus sequence[13] and one could imagine that a mutation at this site is compensated by the surrounding sequence.

In all three operators, the major contacts determined with interference footprinting were contacts to guanosine and thymidine residues, and all focused within a core binding region. Since guanosine is methylated at *N7* in the major groove, one can conclude that CcpN contacts its operator sequences primarily, but not exclusively, through contacts in the major groove, as previously found for many other proteins, e.g. RhaS from *E. coli*.[14] Similarly to transcription factor TyrR from *E. coli*,[15] CcpN contacts its target through a large number of bases. Contacts to the sugar-phosphate backbone make only minor contributions to the CcpN-DNA interaction and, thus, do not seem to play an important role. Most probably, extended contacts to bases relieve the necessity to interact with the sugar phosphate backbone. Interestingly, contacts to the sugar phosphate backbone were found mostly downstream from one of the guanosines that provided one of the main contacts.

The occurrence of two binding sites with different contact strengths within one operator is rather peculiar, as many proteins with two binding sites bind these sites with more or less equal affinity.[16-18] In this regard, however, CcpN shows similarities with PurR,[19,20] whose operators have one strong and one weak binding site, too, although the differences are not as pronounced as in the case of CcpN.

CcpN binding sites are located at different positions at each operator

Previous DNase I footprinting experiments indicated that the binding site distribution is different among the three CcpN-regulated promoters.[11,13] Here, we substantiated these findings and determined the exact borders of the single binding sites using chemical interference footprinting experiments. Figure 5 shows that in all three operators, CcpN binding sites are located at different positions relative to the transcription start site.

At the *srl* operator, site I was found to be centred upstream of the -35 box, around -48, and site II centred around -19. Bases within the -35 box were not contacted by CcpN, and only one base of the -10 box exhibited one less close contact. In contrast, in the *pckA* operator sites I and II partially overlapped the -35 box and completely the -10 box, respectively. The *gapB* operator revealed yet another positioning of the binding sites. Here, binding site I covered the -10 box as does site II in the case of *pckA*, and site II was located downstream from the transcription start site with its centre at position +19.

Diverse distribution of operator sites is not an uncommon feature. Beside transcription factors that show conserved binding site positioning, like CytR from *E. coli*,[21] numerous transcription factors bind to operators that are located at varying positions with regard to the promoter as does CcpN. One example is CcpA, the major factor for carbon catabolite repression in *B. subtilis*, whose binding sites, termed *cre* elements, can be positioned differently relative to the transcription start site: Depending on their regulated gene, they are found at e.g. -33, -3 or +37.[22,23] Interestingly, all these *cre* elements mediate transcriptional repression, although their respective repression mechanism has not yet been elucidated.

Based on the distribution of the CcpN binding sites at the three different promoters, it is tempting to speculate about different repression mechanisms.[24] In the case of *srl*, neither the -35 nor the -10 box are covered or contacted by CcpN. This might allow RNA polymerase to bind simultaneously with CcpN to the *srl* promoter, which would exclude repression by steric hindrance and could result in inhibition of open complex formation as found for, e.g., the MerR repressor of *E. coli*.[25] Another conceivable mechanism is inhibition of promoter clearance, as shown for protein P4 of phage Φ29 at the viral A2c promoter.[26] In contrast, at the *pckA* and *gapB* promoters, inhibition of transcription might occur by steric hindrance of RNAP binding, since at least one binding site of these promoters completely covers the -10 box, as it is the case for the Fur protein from *E. coli* as well as many other transcriptional repressors.[27] Future experiments will focus on the elucidation of the repression mechanism of CcpN at all three operators, for which the ligand modulating CcpN activity[13] still needs to be identified.

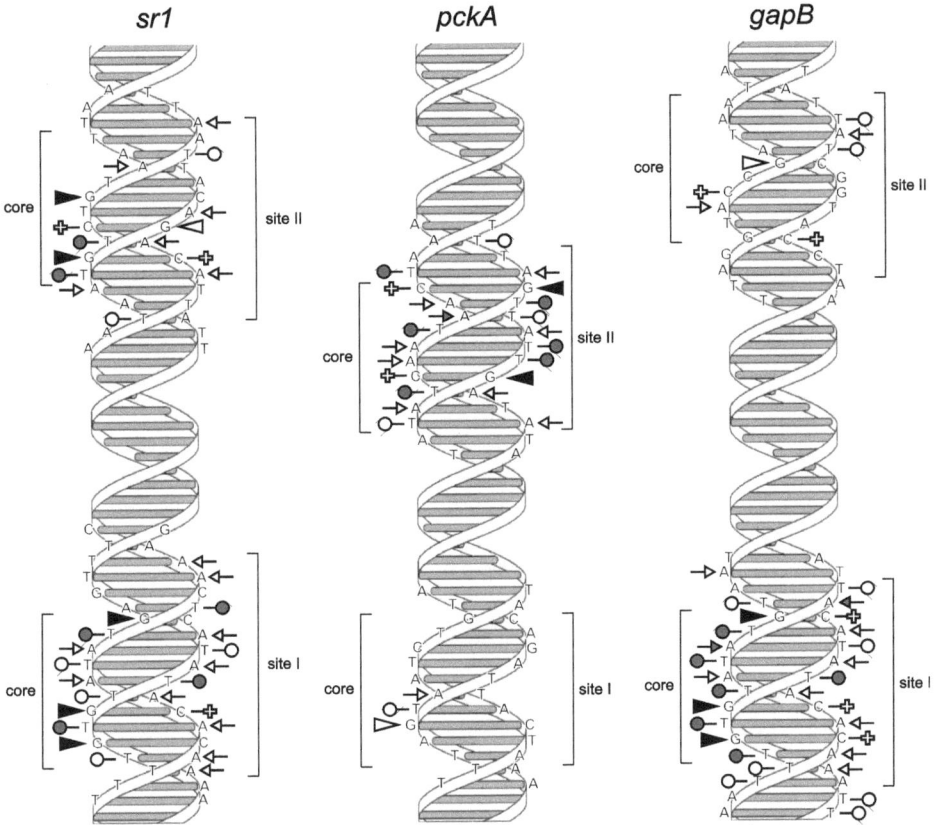

Figure 7: Ribbon model of all CcpN operators

Symbols used to indicate contacts between CcpN and the operator are the same as in Figure 5. The binding sites and the core binding regions are indicated by brackets.

CcpN binds cooperatively to its two binding sites

Our interference footprinting experiments indicated that CcpN contacts its respective binding sites with different strength, especially at the *pckA* and *gapB* promoters (Figure 6). These results were confirmed by EMSAs using oligonucleotides carrying the single binding sites. The determined K_D values varied greatly from as low as 98 nM for the strong site of *gapB* to 4.4 µM for the associated weak site. The same was true for the strong and the weak site of the *pckA* operator, whereas the K_D values for the binding sites of the *sr1* operator did not differ much with 420 nM and 650 nM for site I and II, respectively, due to the only slight differences in contacts between these sites (Figure 6 (a) and Table 1). Surprisingly, the K_D values obtained in EMSAs with DNA fragments containing the whole operators were greatly reduced – up to 340-fold for the weak site of *gapB* – compared to those for the single binding sites and correlated well to what was found in the work by Servant et al.[11] Obviously, two binding sites on one DNA strand dramatically increase the binding efficiency of CcpN.

The determination of the K_D values for the single site in the context of the complete operators confirmed these results. Here, in all three operators both sites were occupied with almost the same efficiency and showed only slight variations in K_D values between the strong and the weak sites. In addition, all K_D values were, partly significantly, decreased. In the *srl* operator, both binding sites showed an almost equal increase in binding affinity, whereas at the *pckA* and *gapB* operators, only the weak binding sites exhibited a significantly lower K_D in the complete operator. Thereby, the affinity of the strong binding sites was mainly unchanged or only slightly increased. This increase in binding affinity leads to an increase in energy gain upon CcpN binding: Binding to two sites that are in close vicinity on one DNA strand is energetically more favourable than binding two separated sites. Furthermore, a change in the shape and slope of the binding curves, resulting from a Hill coefficient h>1, which indicates cooperativity, was observed. All this leads to the conclusion that CcpN apparently binds to its operators in a cooperative way, but this cooperativity is different for the three promoters. While the *srl* operator shows two more or less equal binding sites, the *pckA* and *gapB* operators are composed of one main and one auxiliary site, and binding to the auxiliary site was found to be greatly improved in the presence of the main site. Strong and weak binding sites were also observed for the DeoR repressor-operator system in *B. subtilis*.[28] The DeoR operator consists of one full and one half binding site, but unlike CcpN, DeoR does not bind single sites.

Cooperative binding suggests an interaction between the CcpN molecules bound to the stronger and weaker sites. Conspicuously, the spacer region between the two binding sites differs between the three operators (see Figures 5 and 7). Whereas in the *srl* and *gapB* operator it comprises 3 helical turns, in the *pckA* operator, only two helical turns separate the two binding sites. This indicates that, at least in the case of *srl* and *gapB*, CcpN most likely bends its operator DNA to enable a contact between the two binding entities.

CcpN binding is driven exclusively by a strong binding enthalpy

The determination of binding constants and free binding energy showed that CcpN binding to its operator is unfavourable in terms of entropy change, i.e. entropy decreases upon CcpN-DNA interaction. This effect is overcome by a strong favourable enthalpic contribution, likely due to the numerous and close contacts made to the bases in the operator sequence. This has a clear practical consequence for *B. subtilis*: Since this species tolerates temperatures from as low as 12 °C to as high as 52 °C, a strong binding enthalpy, which is temperature independent, and a low binding entropy change, whose contribution to total binding energy depends on the temperature, ensures that CcpN retains its binding affinity and K_D value for its operators over a large temperature scale.

2.5. Materials and Methods

Preparation of labelled CcpN targets

Oligonucleotides were prepurified by treatment with piperidine for 30 minutes at 90° C to avoid contamination with depurinated DNA resulting from the removal of the protective groups during synthesis. Subsequently, prepurified oligonucleotides were 5' end-labelled with [γ-^{32}P]ATP using T4-polynucleotide kinase (NEB) and purified from 15 % denaturing polyacrylamide gels.[29] Pairwise combinations of labelled and unlabelled oligonucleotides were annealed by incubation at 65 °C for 5 minutes and subsequent slow cooling down to 37 °C. Top and bottom strand of all oligonucleotides carry two G or C residues, respectively, at each end to facilitate correct annealing and to promote additional stability. Labelled double-stranded DNA fragments for the EMSAs with the whole operator sequences were obtained by PCR using the according primer pairs (all oligonucleotides used in this study are summarised in Table 3). The PCR products were purified from an ethidium bromide stained 6 % native polyacrylamide gel and 5' end-labelled as described above. The DNA was then separated from unincoorported [γ-^{32}P]ATP by a sephadex column.

Overexpression and purification of CcpN

A *ccpN* overexpression strain was constructed by cloning a *NcoI/BglII* digested PCR fragment obtained with primers SB673 and SB674 on chromosomal DNA of *B. subtilis* DB104 into the pQE60 *NcoI/BglII* vector (Qiagen). The resulting vector was designated pQGDR. For cloning and subsequent purification of the C-terminally His-tagged protein, *E. coli* strain TG1(REP4) was used. The sequence was confirmed using a Sequenase kit from Amersham Bioscience.

A TG1(REP4, pQGDR) overnight culture grown in TY with 50 µg/ml ampicillin and 25 µg/ml kanamycin was diluted 100-fold, grown for 3 additional hours and induced with 1 mM IPTG. After 2.5 hours, cells were harvested by centrifugation and sonicated 10 minutes in sonication buffer (50 mM sodium phosphate, pH 8.0; 300 mM NaCl; 10 mM imidazole). The supernatant obtained by centrifugation was purified over a Ni-agarose column (Qiagen). The column was washed twice with washing buffer (50 mM sodium phosphate pH 8.0, 300 mM NaCl, 20 mM imidazole), and, afterwards, CcpN was eluted with elution buffer (50 mM sodium phosphate pH 8.0, 300 mM NaCl, 250 mM imidazole). Purification was followed by SDS-polyacrylamide gel electrophoresis. In this way, approximately 80% pure CcpN-His$_5$ was obtained that was stored with 50 % glycerol at -20 °C.

Table 3: Oligonucleotides used in this study

Designation	Sequence	Purpose
SB499	5' GGAAAATGTGTTATACAGTTTGG	*sr1*, site I, upper strand
SB500	5' CCAAACTGTATAACACATTTTCC	*sr1*, site I, lower strand

SB964	5' GGTAAATGTGACATATTAATAGG	*sr1*, site II, upper strand
SB965	5' CCTATTAATATGTCACATTTACC	*sr1*, site II, lower strand
SB962	5' GGAAATAGTATAGACTATTTGGG	*pckA*, site I, upper strand
SB963	5' CCCAAATAGTCTATACTATTTCC	*pckA*, site I, lower strand
SB602	5' GGAATATATGTTATACTAATTGG	*pckA*, site II, upper strand
SB603	5' CCAATTAGTATAACATATATTCC	*pckA*, site II, lower strand
SB598	5' GGTTAATGTGTTATACTAATTGG	*gapB*, site I, upper strand
SB599	5' CCAATTAGTATAACACATTAACC	*gapB*, site I, lower strand
SB960	5' GGAAATTAGGATGGCATAATTGG	*gapB*, site II, upper strand
SB961	5' CCAATTATGCCATCCTAATTTCC	*gapB*, site II, lower strand
SB869	5' GGATATGATGATATGAAATAAAATGTGTTATACAGTTTGTTGTTGACATTTTAAATGTGACATATTAATATAATAACAACAAAAGAAGG	*sr1*, complete operator, upper strand
SB870	5' CCTTCTTTTGTTGTTATTATATTAATATGTCACATTTAAAATGTCAACAACAAACTGTATAACACATTTTATTTCATATCATCATATCC	*sr1*, complete operator, lower strand
SB886	5' GGATGCACACACAAAAAACAATAAATAGTATAGACTATTTGAAAATATATGTTATACTAATTCACAATTAGCAAAACACAAAAAACGGG	*pckA*, complete operator, upper strand
SB887	5' CCCGTTTTTTGTGTTTTGCTAATTGTGAATTAGTATAACATATATTTTCAAATAGTCTATACTATTTATTGTTTTTTGTGTGTGCATCC	*pckA*, complete operator, lower strand
SB894	5' GGTACTGGCGAATTTGTTTTAATGTGTTATACTAATTTTAGATAGTAACAAATTAGGATGGCATAATTGATAAGGGGTGTCCAACATGG	*gapB*, complete operator, upper strand
SB895	5' CCATGTTGGACACCCCTTATCAATTATGCCATCCTAATTTGTTACTATCTAAAATTAGTATAACACATTAAAACAAATTCGCCAGTACC	*gapB*, complete operator, lower strand
SB673	5' GAATTCCCATGGGAAGTACGATCGAACTAAAT	plasmid pQGDR
SB674	5' CTGCAGAGATCTTTATTAGTGATGGTGATGGTGTAGGATTTCATTTTCAGA	plasmid pQGDR
SB342	5' CCCAGGAGAAATTATTACAG	*sr1* downstream primer
SB423	5' TCGAGGATCCAACAAGGTGAATATGATGAT	*sr1* upstream primer

SB1027	5'GAGGGCAGTCAGTGCGGAGC	*gapB* upstream primer
SB1028	5'CAATAAAAAATAAAAAGCATGCGGCTTTAAGCCGCAT GCTTTTTTAGCCACAACCTCTTTGTCGT	*gapB* downstream primer
SB1029	5'AGAGTATCCGCTCAATGAAA	*pckA* upstream primer
SB1030	5'CAATAAAAAATAAAAAGCATGCGGCTTTAAGCCGCAT GCTTTTTTTGTTGTCGCGCGAACAGCAC	*pckA* downstream primer

Oligonucleotides SB673 and SB674 were used to construct plasmid pQGDR. SB342, SB423 and SB1027-SB1030 were used as primers for the amplification of whole operator fragments. All other oligonucleotides were annealed pairwise to create double stranded targets for footprinting experiments and EMSAs.

EMSA and determination of apparent equilibrium dissociation constant KD

Binding reactions were performed in a final volume of 10 µl containing 0.5x TBE buffer, 0.05 g/l herring sperm DNA as non-specific competitor, 1 nM of end-labelled DNA fragment and 5.2 nM to 17.7 µM of CcpN-His$_5$. All CcpN-His$_5$ dilutions were made in storage buffer and the same volume of diluted protein was used in each sample to ensure an equal salt concentration. After incubation at 37° C for 15 min, the reaction mixtures were separated on 6 % (for whole operator DNA fragments) or 8 % (for 23 bp DNA fragments) native polyacrylamide gels run at room temperature for 1 hour at 200 V. Visualisation and quantification of the bands were performed using a Fuji-PhosphorImager and the PCBAS 2.09 quantification software (Raytest). All autoradiograms were made from dried gels. The image data generated by scanning the gel are linear proportionally to the radiation intensity of the sample. The amount of CcpN-DNA complex relative to the CcpN concentration was fitted with the non-linear regression programme Solver (included in Microsoft® Excel) to the following equation:

$$[C] = \frac{[D] \cdot [P]}{K_D + [P]}$$

where [C], [D] and [P] represent total concentrations of formed complex, DNA and protein, respectively, and K_D is the apparent equilibrium dissociation constant.

Methylation interference footprinting

5' end-labelled DNA fragments were modified by DMS as described for the G>A reaction using the Merck oligonucleotide sequencing kit. Modified DNA was subjected to CcpN binding and EMSA as described above. Bound and unbound fractions were separated on a 6 % native polyacrylamide gel and visualised by wet autoradiographic exposure. Bound and unbound DNA was cut out and eluted from the gel by diffusion (elution buffer: 0.15 M NaCl, 50 mM Tris/HCl pH 8.0, 10 mM EDTA pH 8.0), treated with phenol/chloroform and ethanol-precipitated. DNA samples and protein-free DNA as control were depurinated for 15 min at 90 °C, cleaved by piperidine for 30 min at 90 °C, ethanol-precipitated twice, resuspended in formamide loading dye and separated on a 15 % sequencing gel.

Potassium permanganate interference footprinting

5' end-labelled DNA fragments were modified by KMnO$_4$ according to Rubin and Schmid.[30] Modified DNA was subjected to CcpN binding and EMSA as described above. Bound and unbound fractions were separated on a 6 % native polyacrylamide gel and isolated as described above. DNA samples and protein-free DNA as control were cleaved by piperidine for 30 min at 90 °C, ethanol-precipitated twice, resuspended in formamide loading dye and separated on a 15 % sequencing gel.

Hydroxylamine interference footprinting

5' end-labelled DNA fragments were modified by NH$_2$OH according to Rubin and Schmid.[30] Modified DNA was subjected to CcpN binding and EMSA as described above. Bound and unbound fractions were separated on a 6 % native polyacrylamide gel and isolated as described above. DNA samples and protein-free DNA as control were cleaved by piperidine for 30 min at 90 °C, ethanol-precipitated twice, resuspended in formamide loading dye and separated on a 15 % sequencing gel.

Ethylation interference footprinting

5' end labelled DNA fragments were modified by *N*-ethyl-*N*-nitrosourea (Sigma) as described.[31] Modified DNA was subjected to CcpN binding and EMSA as described above. Bound and unbound fractions were separated on a 6 % native polyacrylamide gel and isolated as described above. The DNA was cleaved with 143 mM NaOH at 90 °C for 30 min according to Büning *et al.*[32] Protein-free DNA as control was prepared by NaOH cleavage of an aliquot of the ethylated DNA. After two ethanol precipitations and resuspension in formamide loading dye, the samples were separated on a 15 % sequencing gel.

Densitometric quantification of the footprinting experiments

Band intensities were determined with quantification software (PCBAS 2.09, Raytest) and, afterwards, normalised by dividing them by the total band intensity of the same lane to correct for unequal loading. Data were plotted as logarithm (log) of the ratio of band intensity of bound DNA *versus* band intensity of the unbound DNA for each base position. Negative values were interpreted as interference signals.

Quantitative DNase I footprinting

DNase I footprinting was performed in a final volume of 10 µl containing 0.5x TBE, 6.25 mM MgCl$_2$, 0.05 g/l herring sperm DNA, 1 nM of end-labelled DNA fragment and 8.1 nM to 5.9 µM of CcpN-His$_5$. After incubation at 37 °C for 30 min, the samples were treated with 1 µl of DNase I (Roche, 0.05 U/µl) for 2 min at 37 °C. Two control samples, one without protein, one without DNase I, were treated in parallel. The reaction was stopped by phenol extraction and subsequent ethanol precipitation. The pellets were dissolved in 3 µl formamide loading dye, denatured for 5 min at 90 °C and separated on a 15 % denaturing polyacrylamide gel along with a

Maxam-Gilbert sequencing reaction obtained from the same DNA fragment. The dried gel was analysed by PhosphorImaging. DNA occupancy by CcpN was determined by measuring the band intensity at the binding sites divided by the intensity at an unoccupied part of the DNA.

To ensure that the CcpN-DNA complex is at equilibrium, footprinting experiments with different incubation times prior to DNase I cleavage were carried out. Steady state was reached no later than after 5 min of incubation. To show that DNase I is not able to displace CcpN from its operator, footprinting experiments with different concentration of DNase I were performed. The amount of CcpN-DNA complex relative to the CcpN concentration was fitted with the non-linear regression programme Solver (included in Microsoft® Excel) to the Hill equation:

$$[C] = \frac{[D] \cdot [P]^h}{K_D^h + [P]^h}$$

where [C], [D] [P] and h represent total concentrations of formed complex, DNA, protein and the Hill coefficient, respectively, and K_D is the apparent equilibrium dissociation constant.

Acknowledgements

We would like to thank E. Birch-Hirschfeld (Institut für Virologie, Jena) for synthesising the oligodeoxyribonucleotides. This work was supported by grant BR1552/6-1 from Deutsche Forschungsgemeinschaft to S. B. Andreas Licht is financed by a scholarship from the "Fonds der chemischen Industrie".

2.6. References

1. Steinmetz, M. (1993). Carbohydrate catabolism: pathways, enzymes, genetic regulation, and evolution. In *Bacillus subtilis and other gram-positive bacteria: biochemistry, physiology, and molecular genetics* (Sonenshein, A. L., Hoch, J. A. & Losick, R., eds), pp. 157-170, Am. Soc. Microbiol., Washington, DC.

2. Reizer, J., Bachem, S., Reizer, A., Arnaud, M., Saier, M. H. Jr. & Stülke, J. (1999). Novel phosphotransferase system genes revealed by genome analysis: the complete complement of PTS proteins encoded within the genome of *Bacillus subtilis*. *Microbiology* **145**, 3419-3429.

3. Monod, J. (1942). Recherches sur la Croissance des Cultures Bacteriennes. *Paris: Hermann et Cie.*

4. Chambliss, G. H. (1993). Carbon source mediated catabolite repression. In *Bacillus subtilis and other gram-positive bacteria: biochemistry, physiology, and molecular genetics* (Sonenshein, A. L., Hoch, J. A. & Losick, R., eds), pp. 212-219, Am. Soc. Microbiol., Washington, DC.

5. Emmer, M., deCrombrugghe, B., Pastan, I. & Perlman, R. (1970). Cyclic AMP receptor protein of *E. coli*: Its role in the synthesis of inducible enzymes. *Proc. Natl. Acad. Sci. USA* **66**, 480-487.

6. Eron, L., Arditti, R., Zubay, G., Connaway, S. & Beckwith, J. R. (1971). An adenosine 3':5'-cyclic monophosphate-binding protein that acts on the transcription process. *Proc. Natl. Acad. Sci. USA* **68**, 215-218.

7. Mach, H., Hecker, M. & Mach, F. (1984). Evidence for the presence of cyclic adenosine monophosphate in *Bacillus subtilis*. *FEMS Microbiol. Letters* **22**, 27.

8. Stülke, J. & Hillen, W. (2000). Regulation of carbon catabolism in Bacillus species. *Annu. Rev. Microbiol.* **54**, 849-880.

9. Fillinger, S., Boschi-Muller, S., Azza, S., Dervyn, E., Branlant, G. & Aymerich, S. (2000). Two glyceraldehyde-3-phosphate dehydrogenases with opposite physiological roles in a nonphotosynthetic bacterium. *J. Biol. Chem.* **275**, 14031-14037.

10. Yoshida, K., Kobayashi, K., Miwa, Y., Kang, C. M., Matsunaga, M., Yamaguchi, H., Tojo, S., Yamamoto, M., Nishi, R., Ogasawara, N., Nakayama, T. & Fujita, Y. (2001). Combined transcriptome and proteome analysis as a powerful approach to study genes under glucose repression in *Bacillus subtilis*. *Nucleic Acids Res.* **29**, 683-692.

11. Servant, P., Le Coq, D. & Aymerich, S. (2005). CcpN (YqzB), a novel regulator for CcpA-independent catabolite repression of *Bacillus subtilis* gluconeogenic genes. *Mol. Microbiol.* **55**, 1435-1451.

12. Goldie, H. & Medina, V. (1990). Physical and genetic analysis of the phosphoenolpyruvate carboxykinase (*pckA*) locus from *Escherichia coli* K12. *Mol. Gen. Genet.* **220**, 191-196.

13. Licht, A., Preis, S. & Brantl, S. (2005). Implication of CcpN in the regulation of a novel untranslated RNA (SR1) in *Bacillus subtilis*. *Mol. Microbiol.* **58**, 189-206.

14. Bhende, P. M. & Egan, S. M. (1999). Amino acid-DNA contacts by RhaS: an AraC family transcription activator. *J. Bacteriol.* **181**, 5185-5192.

15. Hwang, J. S., Yang, J. & Pittard, A. J. (1999). Specific contacts between residues in the DNA-binding domain of the TyrR protein and bases in the operator of the *tyrP* gene of *Escherichia coli*. *J. Bacteriol.* **181**, 2338-2345.

16. Steinmetzer, K. & Brantl, S. (1997). Plasmid pIP501 encoded transcriptional repressor CopR binds asymmetrically at two consecutive major grooves of the DNA. *J. Mol. Biol.* **269**, 684-693.

17. Brenowitz, M., Senear, D. F., Shea, M. A. & Ackers, G. K. (1986). "Footprint" titrations yield valid thermodynamic isotherms. *Proc. Natl. Acad. Sci. USA* **83**, 8462-8466.

18. Lewis, M. (2005). The Lac repressor. *C. R. Biol.* **328**, 521-548.

19. Shin, B. S., Stein, A. & Zalkin, H. (1997). Interaction of *Bacillus subtilis* purine repressor with DNA. *J. Bacteriol.* **179**, 7394-7402.

20. Bera, A. K., Zhu, J. & Zalkin, H. (2003). Functional dissection of the *Bacillus subtilis pur* operator site. *J. Bacteriol.* **185**, 4099-4109.

21. Collado-Vides, J., Magasanik, B. & Gralla, J. D. (1991). Control site location and transcriptional regulation in *Escherichia coli. Microbiol. Rev.* **55,** 371-394.
22. Weickert, M. J. & Chambliss, G. H. (1990). Site-directed mutagenesis of a catabolite repression operator sequence in *Bacillus subtilis. Proc. Natl. Acad. Sci. USA* **87,** 6238-6242.
23. Grundy, F. J., Turinsky, A. J. & Henkin, T. M. (1994). Catabolite regulation of *Bacillus subtilis* acetate and acetoin utilization genes by CcpA. *J. Bacteriol.* **176,** 4527-4533.
24. Rojo, F. (2001). Mechanisms of transcriptional repression. *Curr. Opin. Microbiol.* **4,** 145-151.
25. Heltzel, A., Lee, I. W., Totis, P. A. & Summers, A. O. (1990). Activator-dependent preinduction binding of sigma-70 RNA polymerase at the metal-regulated *mer* promoter. *Biochemistry* **29,** 9572-9584.
26. Monsalve, M., Mencia, M., Salas, M. & Rojo, F. (1996). Protein p4 represses phage phi 29 A2c promoter by interacting with the alpha subunit of *Bacillus subtilis* RNA polymerase. *Proc. Natl. Acad. Sci. USA* **93,** 8913-8918.
27. Escolar, L., Perez-Martin, J. & de Lorenzo, V. (1998). Coordinated repression *in vitro* of the divergent *fepA-fes* promoters of *Escherichia coli* by the iron uptake regulation (Fur) protein. *J. Bacteriol.* **180,** 2579-2582.
28. Zeng, X. & Saxild, H. H. (1999). Identification and characterization of a DeoR-specific operator sequence essential for induction of *dra-nupC-pdp* operon expression in *Bacillus subtilis. J. Bacteriol.* **181,** 1719-1727.
29. Sambrook, J., Fritsch, E. F. & Maniatis, T. (1989). Molecular cloning. A laboratory manual. Cold Spring Harbor Laboratory Press, New York.
30. Rubin, C. M. & Schmid, C. W. (1980). Pyrimidine-specific chemical reactions useful for DNA sequencing. *Nucleic Acids Res.* **8,** 4613-4319.
31. Rimphanitchakit, V. & Grindley, N. D. F. (1991). The study of protein-DNA contacts by ethylation interference. In *A laboratory guide to in vitro studies of protein-DNA interactions* (Jost, J. P. and Saluz, H. P., eds), pp. 111-120, Birkhäuser Verlag, Basel.
32. Büning, H., Baeuerle, P. & Zorbas, H. (1995). A new interference footprinting method for analyzing simultaneously protein contacts to phosphate and guanine residues on DNA. *Nucleic Acids Res.* **23,** 1443-1444.

3. Identification of ligands affecting the activity of the transcriptional repressor CcpN from Bacillus subtilis.

Andreas Licht[1]*, Ralph Golbik[2] & Sabine Brantl[1]

[1]AG Bakteriengenetik, Friedrich-Schiller-Universität Jena, D-07743 Germany
[2]Institut für Biotechnologie, Abteilung Mikrobielle Biotechnologie, Martin-Luther-Universität Halle-Wittenberg, D-06120 Halle (Saale), Germany

Published in: *Journal of Molecular Biology*, **380**: 17-30 (2008)

*corresponding author

3.1. Summary

Carbon catabolite repression in *B. subtilis* is mediated primarily by the major regulator CcpA. However, sugar dependent repression of three genes, *sr1* encoding a small nontranslated RNA, and two genes coding for gluconeogenic enzymes, *gapB* and *pckA*, is carried out by the transcriptional repressor CcpN. It has previously been shown that *ccpN* is constitutively expressed, which leads to a constant occupation of all operators with CcpN. Since this would not allow for specific regulation, a ligand is required that modulates CcpN activity. *In vitro* transcription assays demonstrated that CcpN is able to specifically repress transcription to a small extent at the three mentioned promoters in the absence of an activating ligand. Upon testing of several ligands, including nucleotides and glycolysis intermediates, it could be shown that ATP is able to specifically enhance the repressing activity of CcpN, and this effect was more pronounced at a slightly acidic pH. Furthermore, ADP was found to specifically counteract the repressive effect of ATP. CD measurements demonstrated a significant alteration of CcpN structure in the presence of ATP at acidic pH and in the presence of ADP. EMSAs revealed that neither ATP nor ADP altered the affinity of CcpN for its operators. Therefore, we hypothesise that the effect of ligand-bound CcpN on the RNA polymerase might be due to a conformational switch that alters the interaction between the two proteins. Based on these results a working model for CcpN action is discussed.

3.2. Introduction

Most bacteria, among them *Bacillus subtilis*, are able to use a huge variety of nutrients.[1,2] Nonetheless, glucose is the preferred carbon source for most of them.[3] This requires other catabolic pathways to be shut down in the presence of glucose to maximise their energy yield. This process of catabolite repression in *B. subtilis* is mediated mainly by the concerted action of CcpA and HPr-Ser46-P, which can interact to form a transcriptional regulator.[4] Though the majority of genes involved in carbon metabolism are regulated by the CcpA/HPr system, at least three genes, *gapB*, *pckA* and *sr1*, are downregulated in the presence of glucose by an alternative transcriptional repressor named CcpN (control catabolite protein of gluconeogenic genes) which exerts its function under glycolytic conditions.[5,6,7] GapB and pckA encode enzymes that are exclusively active during gluconeogenesis,[5,8] while *sr1* codes for a small untranslated RNA, which has been identified by a systematic search for small RNAs within intergenic regions of the *B. subtilis* genome.[9] The *sr1* gene was also found to be expressed during gluconeogenesis, but repressed under glycolytic conditions. Its gene product, SR1, inhibits translation initiation of *ahrC* mRNA, encoding a transcriptional activator of the arginine catabolic operons, by a novel mechanism. Seven regions of complementarity between SR1 and ahrC have been found, designated A to G. Upon SR1/ahrC interaction, structural alterations are induced between the ahrC ribosome binding site and region G located 90 nt downstream from it. These structural alterations prevent the binding of the 30S ribosomal subunit.[10,11]

The *ccpN* gene forms a bicistronic operon with the *yqfL* gene, whose function is not yet fully clear. This operon is not autoregulated, but constitutively expressed under both glycolytic and gluconeogenic conditions.[7] Homologues of CcpN have been found in the genomes of other Bacilli, e.g. *B. halodurans, B. cereus, B. anthracis* and *Geobacillus stearothermophilus*, and in different firmicutes.[7]

Recent investigations have demonstrated that CcpN occupies two distinct binding sites at each of the three regulated promoters. The position of the operator sites with respect to the transcriptional start site varies depending on the promoter, but in each case one of these sites is contacted more efficiently than the other one. However, it has been shown that both binding sites are bound with equal affinity when located in close vicinity, since CcpN binds its half-sites in a cooperative manner.[12]

The aim of the present work was to identify the ligands that modulate the activity of CcpN. EMSAs (electrophoretic mobility shift assays) demonstrated that none of the investigated potential ligands altered the affinity of CcpN to its operator. Therefore, *in vitro* transcription assays with native *B. subtilis* RNA polymerase were used as an alternative method to investigate the influence of various substances on the repression activity of CcpN. These assays showed a specific increase in repression activity in the presence of high concentrations of ATP and at low pH, whereas high concentrations of ADP were able to counteract the effect of ATP. Furthermore, CD measurements have been performed that revealed a substantial ATP-induced alteration of CcpN secondary structure. The combination of these data sets allowed to develop a new working model on the mechanism of action of CcpN.

3.3. Results

In vitro transcription experiments were performed with *B. subtilis* crude extracts from a CcpN knockout strain (DB104 *ccpN::cat*)[9] that were filtrated through a Millipore column (molecular weight cutoff: 100,000 Da). This allows the separation of the RNAP holoenzyme from smaller proteins, but retains any RNA polymerase associated factors. It has been confirmed previously that RNAP purified this way yields the same results as His-tagged *B. subtilis* RNAP purified according to the protocol of Fujita *et al.*,[13] and as native *B. subtilis* RNAP prepared according to the protocol of Sogo *et al.* (data not shown).[14]

To ensure that *yqfL* has no effect on the metabolic regulation of the *srl* gene, a *ccpN/yqfL* double knockout strain was complemented with a plasmid carrying the *ccpN* gene under control of p_{Spac}. Since the *ccpN* gene itself is not regulated,[7] this strain – after proper induction – behaves as a *yqfL* single knockout strain. Northern blot analyses revealed that this strain shows a response to glucose like the wild type strain, although with a slightly reduced general *srl* transcription level (Figure S1). This corresponds perfectly to the findings of Servant *et al.*,[7] who observed the same effects when investigating the influence of YqfL on *gapB* and *pckA* regulation. Therefore all effects observed below can be attributed to the action of CcpN alone.

CcpN is able to specifically repress transcription at the *srl*, *gapB* and *pckA* promoters

To determine whether CcpN *per se* is able to repress transcription without the addition of a ligand, linear DNA molecules carrying the *srl*, *gapB*, *pckA* or *RNAII* and *RNAIII* promoter, respectively, were incubated with increasing concentrations of CcpN and used as a template for an

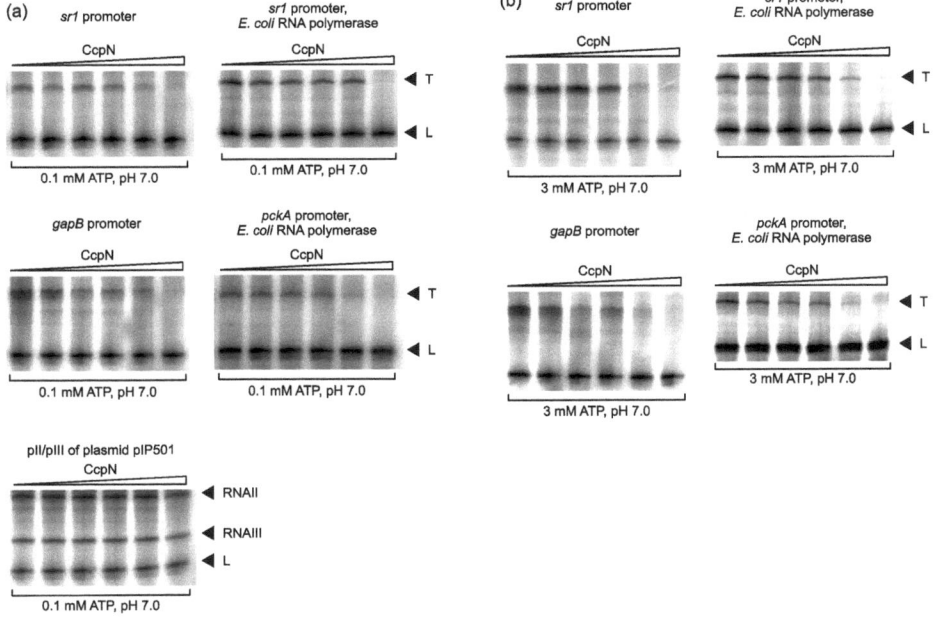

Figure 1: Effect of CcpN and ATP on *in vitro* transcription

In vitro transcription assay at pH 7.0 using 100 nM of a DNA fragment containing the promoter of *sr1*, *gapB*, *pckA* or *RNAII/III* at (a) 0.1 mM ATP or (b) 3 mM ATP. 100 nM of *B. subtilis* RNAP were used in each reaction. CcpN concentration from left to right was 0 nM, 78 nM, 156 nM, 313 nM, 625 nM and 1.25 µM, respectively. Where indicated, 50 nM of *E. coli* RNA polymerase were used. A radioactively labelled 89 nt DNA fragment served as a loading control to ensure equal amounts of the reaction being loaded onto each lane. The bands corresponding to the transcript (T) and to the loading control (L) are indicated. The autoradiograms of the gels are shown.

in vitro transcription reaction. *In vitro* transcriptions were performed with *B. subtilis* RNAP for the *sr1* and *gapB* promoters and - as a negative control - for promoters pII and pIII of streptococcal plasmid pIP510, controlling transcription of RNAII and RNAIII,[15] respectively. Since in the case of *pckA*, *B. subtilis* RNAP yielded only very faint bands, *E. coli* RNAP was used instead. To ensure that the results obtained with *E. coli* RNAP were comparable to those obtained with *B. subtilis* RNAP, all key experiments with the *sr1* promoter were performed with both polymerases. Figure 1 (a) shows the results and Figure 6 summarises all *in vitro* transcription experiments for better clarity. Once the CcpN concentration exceeded a certain threshold, all the promoters that are subject to regulation by CcpN *in vivo* revealed reduced transcription. By contrast, promoters pII and pIII, which are not subject to regulation by any *B. subtilis* protein, were not affected by CcpN even at very high concentrations. The observation that the *sr1*, *gapB* and *pckA* promoters are repressed by

CcpN even in the absence of an added ligand corresponds very well to the observations made by Servant et al.,[7] who reported a significant derepression of the *gapB* and the *pckA* gene in a *ccpN* knockout strain.

LacZ fusions show that different glycolysis mutants influence repression by CcpN

Since the presence of glucose in the medium influences CcpN activity, we constructed transcriptional *srl-lacZ* fusions to investigate whether intermediates of the glycolytic pathway affect CcpN. These constructs were integrated into the chromosome of *B. subtilis* strains that bear mutations in different glycolytic genes, thus interrupting glycolysis at certain steps. Strain QB5331 harbours a knockout of glucose-6-phosphate isomerase and strain SU22 a mutation in the glyceraldehyde-3-phosphate-dehydrogenase gene.[16] Both strains showed growth curves similar to the wild type (data not shown). β-galactosidase measurements, summarised in Table 1, showed that in the wild-type strain *srl* expression was repressed ≈ 37-fold. Strain SU22 exerted a CcpN-mediated repression of factor 33, which is still significant but not as strong as the wild type. By contrast, strain QB5331 suffered from a severe lack in the ability to respond to CcpN in the presence of glucose in the medium, as it only showed a repression factor of 3.5. These results might imply that one of the glycolysis intermediates between glucose-6-phosphate and 1,3-bisphosphoglycerate is the ligand of CcpN. This hypothesis was surveyed in *in vitro* transcription assays.

Table 1: Results of β-galactosidase measurements

Strain	Mutation	- Glucose (miller units)	+ Glucose (miller units)	Repression factor
DB104	none	890 (± 63)	24 (± 4)	37
SU22	*gapA*	523 (± 5)	16 (±5)	33
QB5331	*pgi*	359 (± 89)	104 (± 14)	3.5

Summary of β-galactosidase measurements with wild-type *B. subtilis* and different strains with mutated glycolysis genes. Denoted mutations refer only to mutations in genes of the glycolytic pathway. Cultures were grow in SP medium to an OD_{600} of 2.0 (early stationary phase). Data are averaged from three independent experiments.

Carbon catabolism intermediates do not affect CcpN-mediated repression

To test whether certain molecules, especially glycolysis intermediates, affect the repression effect of CcpN, *in vitro* transcription assays were performed in the presence of a variety of substances, including nucleotides and carbon catabolism intermediates. Since some intermediates are not commercially available and others are present at very low concentrations *in vivo*,[17,18] only certain compounds were tested. A complete list of all tested substances can be found in Table 2. Since the experiments presented in Figure 1 demonstrated that the three promoters respond to CcpN in the same manner, only the *srl* promoter was used as a model promoter for this screening. Figure

2 shows the results of these experiments. Of all tested nucleotides, only ATP had an effect on the transcription level. It increased transcription efficiency by a factor of five, but this effect was not related to the presence of CcpN. This increase in the presence of 3 mM ATP can be explained by the increase in stability of the open complex, since an A is the first nucleotide of all three newly synthesised transcripts. Based on these results, all other substances were tested in the presence of high ATP concentrations to ensure reliable detection of the transcript. As can be seen in Figure 2, the presence of glyceraldehyde-3-phosphate led to a significant decrease in transcription efficiency only in the presence of CcpN. However, closer inspection revealed that the glyceraldehyde-3-phosphate solution was acidic, causing the pH of the *in vitro* transcription buffer to drop from 7.0 to 6.5. Tests performed with neutralised glyceraldehyde-3-phosphate at low ATP concentration, at neutral and acidic pH and in the presence or absence of CcpN showed no effect at all, which attributes the specific repression effect to the acidic pH value (data not shown).

Table 2: List of all tested putative ligands in *in vitro* transcription

Substance	Relative transcription	Substance	Relative transcription
control	1.0 x	FAD	1.2 x
ATP	1.0 x	SAM	1.0 x
CTP	0.8 x	citrate	0.9 x
AMP	0.9 x	succinate	1.1 x
ADP	1.0 x	glucose	1.2 x
dAMP	1.3 x	pyruvate	0.9 x
dATP	0.8 x	glyceraldehyde-3-phosphate (acidic)	0.2 x
adenosine	0.8 x		
GMP	0.8 x	glyceraldehyde-3-phosphate (neutral)	0.9 x
GDP	1.3 x		
GTP	1.2 x	fructose-1,6-bisphosphate	1.1 x
dGTP	0.7 x	phosphoenolpyruvate	0.8 x
CMP	0.7 x	glutamate	1.1 x
dCMP	0.9 x	L-arginine	1.1 x
UTP	1.0 x	L-methionine	1.0 x
dTTP	1.3 x	L-lysine	1.2 x
NADH	1.0 x	dihydroxyacetonephosphate	1.0 x
NADPH	1.1 x	2-phosphoglycerate	0.9 x

Summary of all substances investigated in *in vitro* transcription. Relative transcription shows the amount of transcript at 625 mM CcpN divided by the amount of transcript in the absence of CcpN. All substances were applied at 1 mM final concentration. Data are averaged from 3 independent experiments.

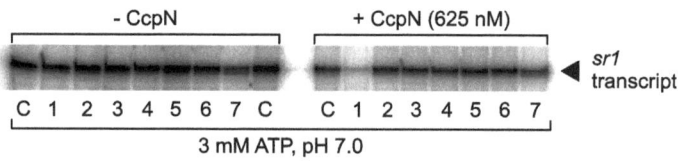

Figure 2: Investigation of putative effectors

In vitro transcription assay at pH 7.0 using 100 nM of a DNA fragment containing the *sr1* promoter. 100 nM of *B. subtilis* RNAP were used in each reaction. All investigated substances were applied at 1 mM final concentration, C, control; 1, glyceraldehyde-3-phosphate (free acid); 2, phosphoenolpyruvate; 3, dATP; 4, pyruvate; 5, citrate; 6, fructose-1,6-bisphosphate; 7, GDP. Table 2 shows a summary of all tested substances. The autoradiogramm of the gel is shown.

ATP specifically enhances CcpN mediated repression at the three promoters

The search for conserved domains in the CcpN sequence revealed, beside the DNA binding domain, a pair of CBS domains.[7] These domains can be found in a variety of proteins in all three kingdoms of life and have been shown to exert different functions, like binding of adenine nucleotides,[19] formation of an oligomerisation interface or parts of an ion transport channel.[20,21] Since binding of ATP or other adenine nucleotides would be very feasible in the case of CcpN, as it reflects the metabolic state of the cell, a series of experiments in the presence of ATP were performed. Since the results obtained at constant CcpN concentrations did not show a specific effect of ATP (Figure 2), the CcpN concentration was varied. As can be seen in Figure 1 (b), the presence of 3 mM ATP decreased the minimal inhibitory concentration of CcpN by approximately factor two at all three promoters. Regarding the efficient expression of these three genes *in vivo*, the effect was considerably smaller than expected.[7,9] Therefore, it seemed that another ligand is required for efficient repression.

Acidic pH value is the second requirement for CcpN-dependent repression

As shown in Figure 2, low pH value in the presence of high ATP concentration led to a strong and specific repression of transcription by CcpN. To examine whether low pH value alone would be sufficient to induce CcpN dependent repression, *in vitro* transcription experiments were performed at constant CcpN concentration in the presence of 0.1 mM ATP while pH was decreased from pH 7.2 to 6.5 (Figure 3 (a)). Alternatively, the effect of increasing CcpN concentration at pH 6.5 and 0.1 mM ATP was investigated (data not shown). Neither of these combinations showed any specific repression at all, implying that low pH is necessary, but alone not sufficient for CcpN activity. A second set of experiments, using the same combinations of pH value and CcpN concentration, but performed at 3 mM ATP, showed a strong specific repression effect (Figures 3 (b) and 4 (a-c)). This indicates that a combination of ATP and low pH is required to unfold the full repression capability of CcpN.

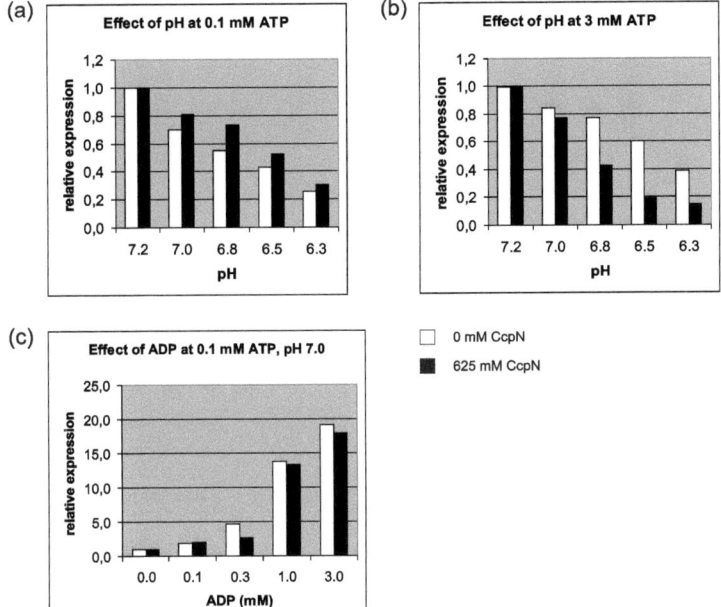

Figure 3: Effect of ATP, low pH and ADP on *in vitro* transcription

In vitro transcription using 100 nM of a DNA fragment and 100 nM of *B. subtilis* RNAP in each reaction. The pH was adjusted with HCl.

(a) Column diagrams of *in vitro* transcription assays at different pH values using the *srl* promoter in the presence of 0.1 mM ATP without or with 625 nM CcpN, as indicated. The relative transcript levels, normalised at pH 7.2, in presence and absence of CcpN are shown.

(b) Column diagrams of *in vitro* transcription assays at different pH values using the *srl* promoter in the presence of 3 mM ATP without or with 625 nM CcpN, as indicated. The relative transcript levels, normalised at pH 7.2, in presence and absence of CcpN are shown.

(c) Column diagrams of *in vitro* transcription assay at different ADP concentrations using the *srl* promoter in the presence of 0.1 mM ATP without or with 625 nM CcpN, as indicated. The relative transcript levels, normalised at 0 mM ADP, in presence and absence of CcpN are shown.

ADP can specifically counteract the effect of ATP to relieve repression by CcpN

Recently, the crystal structure of the regulatory domain of CcpN was solved in the group of N. Declerck who showed that CcpN is, besides binding ATP, also able to bind ADP (Chaix *et al.*, manuscript in preparation). Inspired by this finding, we tested whether ADP had any influence on the repression activity of CcpN in *in vitro* transcription assays. Figure 3 (c) shows that ADP alone

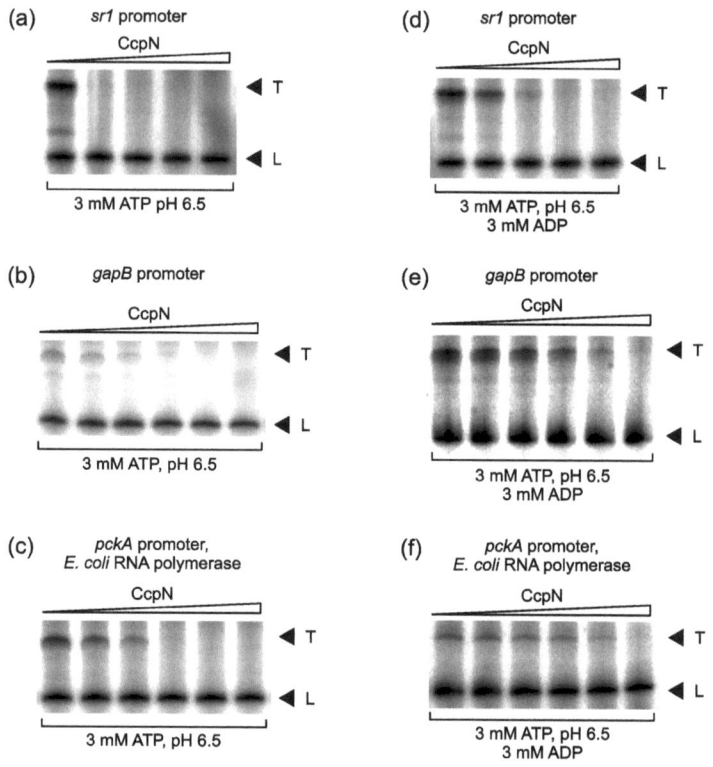

Figure 4: Effect of ADP on *in vitro* transcription

In vitro transcription assay using 100 nM of a DNA fragment and 100 nM of *B. subtilis* RNAP in each reaction. Where indicated, 50 nM of *E. coli* RNA polymerase were used. The loading control was as in Figure 1. The autoradiograms of the gels are shown.

(a) *In vitro* transcription assay at pH 6.5 using the *sr1* promoter in the presence of 3 mM ATP with increasing concentrations of CcpN (0 nM; 313 nM; 625 nM 1.25 µM and 2.5 µM).

(b) *In vitro* transcription assay at pH 6.5 using the *gapB* promoter in the presence of 3 mM ATP with increasing concentrations of CcpN (0 nM; 78 nM; 156 nM; 313 nM; 625 nM and 1.25 µM).

(c) *In vitro* transcription assay at pH 6.5 using the *pckA* promoter in the presence of 3 mM ATP with increasing concentrations of CcpN, as in (b).

(d) *In vitro* transcription assay at pH 6.5 using the *sr1* promoter in the presence of 3 mM ATP and 3 mM ADP with increasing concentrations of CcpN, as in (a).

(e) *In vitro* transcription assay at pH 6.5 using the *gapB* promoter in the presence of 3 mM ATP and 3 mM ADP with increasing concentrations of CcpN as in (b).

(f) *In vitro* transcription assay at pH 6.5 using the *pckA* promoter in the presence of 3 mM ATP and 3 mM ADP with increasing concentrations of CcpN as in (b).

in addition to 0.1 mM ATP did neither increase nor decrease CcpN-mediated repression. However, when equimolar concentrations of ADP were added to an *in vitro* transcription reaction performed with 3 mM ATP at pH 6.5, ADP was capable to completely counteract the repression enhancing effect of ATP (Figure 4 (d-f)).

CcpN mutated in a CBS domain loses its ability to respond to ATP or ADP

To examine whether CcpN with a mutation in its nucleotide binding domain retains its ability to respond to the two nucleotides, a mutant version of CcpN was investigated in *in vitro* transcription: K127A. In this mutant, a conserved amino acid within one of the two CBS domains was replaced by an alanine. The mutant was a kind gift by Stéphane Aymerich and will be published elsewhere (Chaix *et al.*, manuscript in preparation). Figure 6 shows that this mutant exerted the same kind of basic repression that could be observed with the wild type (Figure 1), but lacked the ability to respond to ATP or ADP. According to Stéphane Aymerich (Chaix *et al.*, manuscript in preparation), this mutant is not able to repress *gapB* or *pckA* transcription *in vivo*.

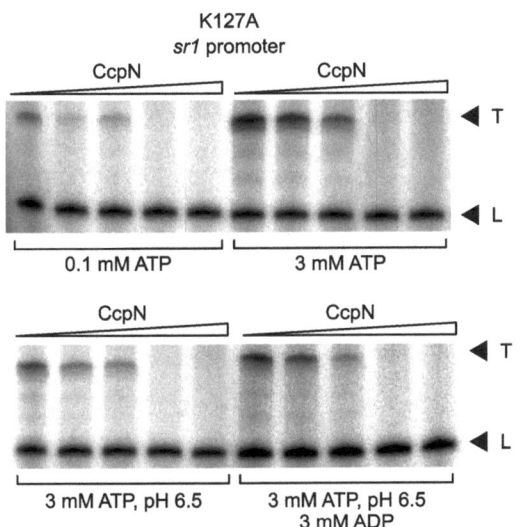

Figure 5: Investigation of the K127A mutant in *in vitro* transcription assays

In vitro transcription assay at the *sr1* promoter using 100 nM of a DNA fragment and 100 nM of *B. subtilis* RNAP in each reaction. CcpN protein with mutation K127A was added at concentrations: 0 nM; 313 nM; 625 nM 1.25 µM and 2.5 µM. Reaction conditions are specified under the gel. The loading control was as in Figure 1. The autoradiograms of the gels are shown.

Repression conditions do not change the affinity of CcpN for DNA

Since ATP and low pH value have a strong effect on CcpN activity, we wanted to analyse whether they affect the affinity of CcpN for its operator sequence. To this end, a double-stranded DNA fragment harbouring the *sr1* operator region was incubated with increasing concentrations of CcpN and subjected to an EMSA. This reaction was performed under non-repressive conditions, in the presence of 3 mM ATP or in the presence of 3 mM ATP and a pH of 6.5 (Figure 7). To ensure that the conditions did not change during electrophoresis, both ATP and low pH, where applicable, were also present in the gel and in the running buffer. However, none of the repression conditions did affect the affinity of CcpN to its operators. This suggests that the specific repression of CcpN induced by ATP and an acidic pH shift is not based on an increased affinity to the promoter.

(a)

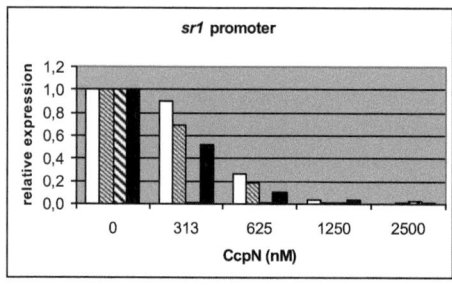

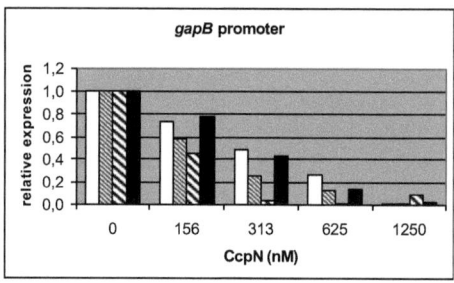

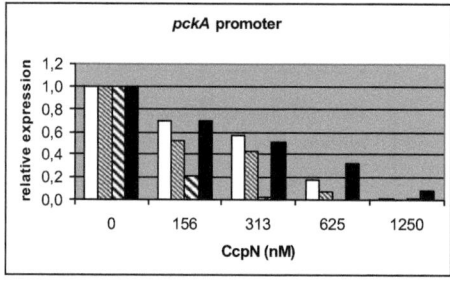

(b)

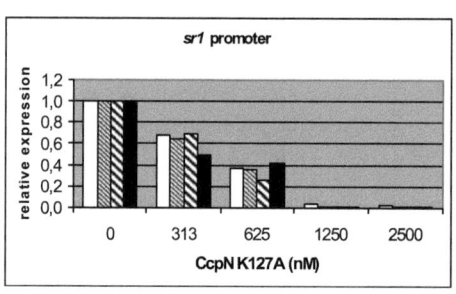

Figure 6: Overview of in vitro transcriptions.

(a) Summary of the effects of ATP, low pH and ADP on CcpN activity at the *sr1*, *gapB* or *pckA* promoter. Reaction conditions are indicated in the inset. Transcription levels have been normalised at 0 mM CcpN.

(b) Summary of the effects of ATP, low pH and ADP on CcpN activity at the sr1, gapB or pckA promoter using CcpN K127A.

CD measurements reveal an influence of ATP and ADP on the protein structure

Circular dichroism experiments were used to detect an influence of ATP or ADP on the secondary structure of CcpN. The far UV CD spectra of the protein in the presence and absence of ATP or ADP at neutral or acidic pH are presented in Figure 8. Without addition of a ligand, the protein displayed two negative extrema near 208 and 222 nm that indicate the presence of α-helical structures. At neutral pH, the spectrum did not alter significantly after addition of ATP. However, when ATP was added at pH 6.5, a substantial decrease in the α-helical content could be observed. This corresponds to the finding that only at acidic pH, ATP was able to increase the repression efficiency of CcpN. On addition of increasing concentrations of ADP, a change in the CD spectrum of CcpN could be observed, too, but not as pronounced as in the case of ATP. Interestingly, the ADP effect did not seem to depend on pH, as it was almost the same at neutral and acidic conditions.

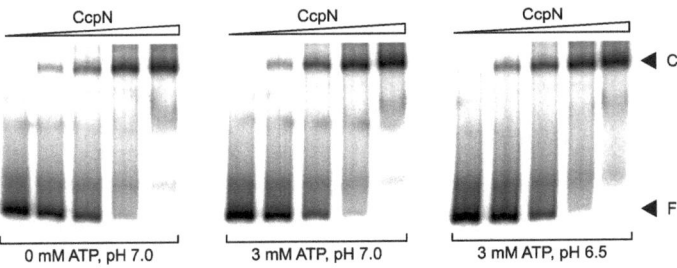

Figure 7: EMSA with CcpN under different conditions

EMSAs of double-stranded 89 bp DNA fragments containing the *srl* operators. The DNA was incubated with increasing concentrations of purified CcpN-His$_5$ (CcpN concentration from left to right: 0 nM; 156 nM; 313 nM; 625 nM and 1.25 µM). Specific reaction conditions are denoted under each gel. The autoradiograms of the gels are shown.

3.4. Discussion

ATP and acidic pH were identified as the two factors required for the full repression capability of CcpN

In this study, we present the identification of ligands that are necessary for CcpN to work as an efficient repressor, as well as an investigation of ligand-protein interaction. It has previously been shown that the CcpN gene is not regulated,[7] which results in a constant concentration of CcpN in the cell under both glycolytic and gluconeogenic conditions. Since CcpN-mediated repression is only required during glycolysis, a ligand is necessary to modulate its activity according to the current metabolic state of the cell.

In vitro transcription assays demonstrated that CcpN is able to exert a semi-specific repression at the three known CcpN-regulated promoters, p_{srl}, p_{gapB} and p_{pckA} without any ligand. By contrast, control promoters, were not affected by CcpN. This finding corresponds very well to the observation of a rather significant derepression of p_{pckA} in a *ccpN* knockout strain.[7] Furthermore, EMSAs performed with *B. subtilis* crude protein extracts revealed a significant amount of bound DNA,[9] which might result from a high CcpN concentration in the cell. Both of these findings provide evidence that the CcpN-regulated promoters are constantly occupied by CcpN and, therefore, partly repressed in *B. subtilis*. Transcriptional regulators that are constitutively bound to their operators are not uncommon. One example is the ResD protein from *B. subtilis*, which induces the *yclJK* operon under oxygen limitation and constantly occupies a single binding site in the

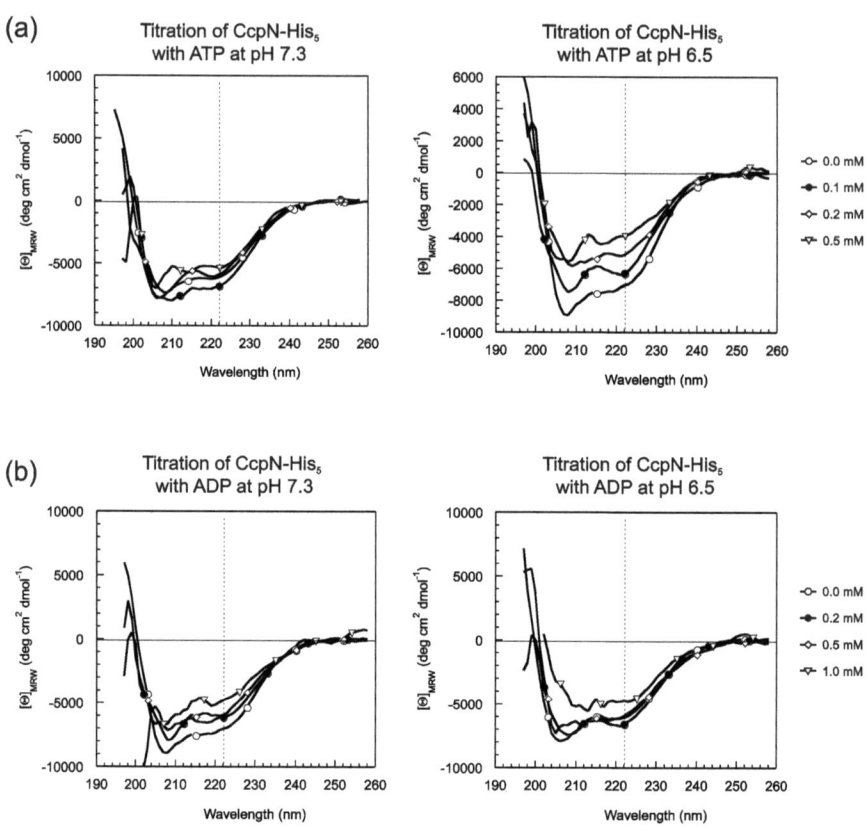

Figure 8: Far-UV CD spectra of CcpN with ligands

Circular dichroism spectrum of CcpN with increasing concentrations of ATP (a) or ADP (b) at neutral or acidic pH. The experiments were performed at 37 °C with 0.108 g/l protein.

promoter region of this operon.[23] A similar situation can be found with the *E. coli* protein NorR, which acts as a transcriptional activator of the detoxification operon *norVW*.[24] However, none of these proteins exerts a constant repression or activation upon their regulated operon, as it has been observed in the case of CcpN. It is not unusual that a transcriptional repressor knockout strain shows a slight derepression of its target gene. BzdR from *Azoarcus sp.*, for example, a repressor of the anaerobic catabolism operon for benzoate, relieves its constitutive repression by a factor of ≈ 1.3 in the *bzdR* knockout strain.[25] Interestingly, an unspecific repression by a factor of 5.4, as exerted by CcpN, is rather peculiar and its biological relevance remains to be elucidated.

β-galactosidase measurements of p_{sr1}-*lacZ* fusions integrated into different strains with mutations in glycolytic genes revealed that a mutation in the *gapA* gene hardly affects CcpN-mediated repression, while a mutation in the *pgi* gene decreases repression efficiency significantly. This led to the assumption that the wanted ligand is an intermediate in this part of the pathway, namely fructose-6-phosphate, fructose-1,6-bisphosphate, glyceraldehyde-3-phosphate or dihydroxyacetone-phosphate. However, investigation of these substances in *in vitro* transcription did not show a specific CcpN-related effect.

It has previously been shown that CBS domains, of which two are found in CcpN, are able to bind to the adenine part of nucleotides and nucleosides.[19] Surprisingly, ATP did not result in an enhanced CcpN-repression in *in vitro* transcription assays with constant CcpN concentration. By contrast, when the CcpN concentration was varied, a small but reproducible effect of ATP was visible: a 2-fold reduction of the minimal inhibitory concentration of CcpN at all three investigated promoters. This amount of repression was much smaller than anticipated from the *lacZ* fusions. This implies that a second factor is required for efficient repression by CcpN. Based on the results obtained with acidic GA3P, a combination of low pH and high ATP concentrations demonstrated that both of these two effectors are required to achieve full CcpN-mediated repression. The dependence on pH is not exceptional, as it is known that a specific pH is required for the correct function of many proteins, among them ion transporters and especially proteases.[26,27] It is noteworthy, however, that a pH sensing function has only been reported for one transcription factor to date, NikR from *E. coli*, whose sensitivity to nickel is dependent on the current pH in the cell.[28] In the case of *B. subtilis* CcpN, the drop in pH in the cell might result from an accumulation of acetate as a final product of the carbon overflow mechanism,[29] which would fit well into the observed regulation performed by CcpN. If excess glucose is available, the citric acid cycle is shut down leading to an accumulation of acetate, which is excreted afterwards, and a slight acidification of the cell.[30] This, however, does not explain the repression effect observed in the □-galactosidase measurements with strain SU22. Since none of the investigated substances showed any significant effect in *in vitro* transcription assays, the observations made with this strain might just be an artefact. It could also be possible that one of these substances exerts its effect via a hitherto unknown protein and indeed enhances CcpN-mediated repression *in vivo*.

CD measurements performed with purified CcpN-His$_5$ in the presence of ATP or ADP at neutral or acidic pH have strengthened the results of the *in vitro* transcription assays. Obviously, ATP binding to CcpN results in an induced fit mechanism, as significant structural changes occur when increasing concentrations of ATP are present. Such induced fit mechanisms are relatively common for ligand-binding proteins, because they are necessary for their regulatory activity. Examples include the multidrug-binding transcriptional repressor QacR from *Staphylococcus aureus*,[31] a wide range of metabolic enzymes, or the human monoamine oxidase, where structural changes have also been detected by CD spectroscopy.[32]

ADP is able to counteract the effect of ATP and HCl

We demonstrated that ADP is - at equimolar concentrations - able to specifically counteract the effect of ATP. CD measurements reinforced these findings, although the observed effect is not as strong as the effect caused by ATP (Figure 7). Soga et al.[17,18] have measured the intracellular concentration of metabolites and nucleotides and have shown that, while there is less ADP than ATP in exponentially growing cells, the ADP concentration exceeds the ATP concentration significantly in cells that have entered the stationary phase. In addition to this, it is generally accepted that there is a sharp drop in intracellular ATP concentration upon glucose limitation, ultimately leading to the activation of the RsbW/RsbV system of cellular stress response.[33,34] Furthermore, it has been demonstrated that CBS domains are able to bind ADP as well as ATP,[19] which corresponds very well to our findings. It is absolutely feasible that ADP, once its concentration is high enough, replaces ATP in the binding pocket which leads to structural rearrangements that ultimately result in a relief of CcpN-repression. Such counterregulation can often be observed with enzymes that have to act differently in the presence of certain signal molecules or metabolites, like aspartate transcarbamoylase from *E. coli* that is stimulated by ATP and inhibited by CTP.[35] However, transcriptions factors are mostly not counterregulated but have just one ligand that turns them "on" or "off", e.g. BzdR and its ligand Benzoyl-CoA from *Azoarcos sp.*[25] One exception, besides CcpN, is GltC from *B. subtilis*, that is activated by α-ketoglutarate and repressed by glutamate,[36] making these proteins in this respect a peculiarity. However, there are also some differences, as GltC is also regulated by RocG *in vivo*,[37] while YqfL does not influence repression by CcpN.

CcpN with a mutation in a crucial residue can no longer exert its function

We have examined a mutation in the CcpN protein in a conserved residue within one of the CBS domains. Aymerich and Declerck (Chaix *et al.*, manuscript in preparation) showed that this protein is not active *in vivo* anymore. However, we observed an unspecific CcpN-mediated repression as in the wild type case. This finding and the fact that the mutant is able to bind to their operator sequence like the wild-type protein imply that the mutation did not affect general DNA binding affinity. Furthermore, Aymerich and Declerck confirmed that this mutant has the same structure as the wild-type (Chaix *et al.*, manuscript in preparation). Interestingly, this mutant does not respond to ATP. According to Aymerich and Declerck, mutant K127A is no longer able to bind ATP or ADP. Consequently, CcpN mutant K127A is unable to perform specific repression because of the lack of ATP binding ability.

New working model on CcpN action

It has been shown that ATP and low pH are specific effectors of CcpN in *in vitro* transcription, but they are not able to alter the binding affinity of CcpN to its operator sequence, as revealed by EMSA. Transcriptional regulators that constantly occupy their operators and share this feature with CcpN include NorR from *E. coli* or ResD, part of a two component system from *B. subtilis*.[23,24] Such proteins usually operate through alterations in structure, induced by a ligand or another activating signal, like phosphorylation in the case of ResD. Our CD data clearly demonstrated that at acidic pH ATP induces significant structural rearrangements in CcpN and, therefore, strongly support this hypothesis. However, what is the mechanism of CcpN action? Three

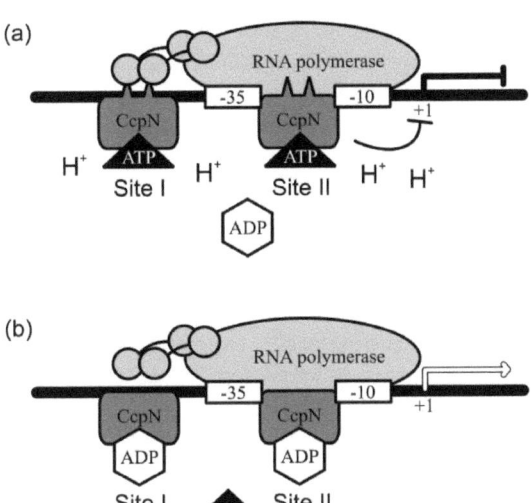

Figure 9: Working model on the mechanism of CcpN-mediated repression

Current working model showing the effectors of CcpN. Binding sites I and II are indicated, 12 C- and N-terminal domain of the RNAP α-subunits are shown as spheres. (a) High glucose concentration in the medium, glycolytic conditions: ADP is only present at very low concentrations and, due to extensive glycolysis, acetate accumulates and acidifies the cell. This allows ATP to bind to CcpN, that now efficiently represses transcription. (b) Low glucose concentration in the medium, gluconeogenic conditions: ADP is present at very high concentrations and able to replace ATP in the nucleotide binding pocket. In addition to this, the citric acid cycle is no longer repressed, no acetate accumulates and the pH turns back to neutral. CcpN is no longer able to repress transcription specifically.

main mechanisms can be postulated, and a general model is shown in Figure 9: CcpN constantly occupies its operators and exerts a certain level of permanent basic repression. It is feasible that CcpN, upon ATP binding, alters its structure in a way that it occupies more space at the promoter region and replaces RNA polymerase, resulting in a classical steric hindrance mechanism, like reported for the Fur protein from *E. coli*.[38] A second possibility would be that ATP-bound CcpN interacts with RNA polymerase. This interaction could influence several phases of transcription initiation. One could imagine that an interaction between CcpN and RNA polymerase inhibits open complex formation, as has been reported for the MerR repressor of *E. coli*.[39] A third alternative mechanism would be the inhibition of promoter clearance, as shown for protein P4 of phage φ29 at the viral A2c promoter.[40] It is interesting to note that, depending on the binding site, CcpN might contact different parts of the RNA polymerase holoenzyme. CcpN bound at site I could contact the C-terminal domain of the α-subunit, while CcpN at site II might form contacts to the sigma factor. Interestingly, it is not yet clear whether two contacts to the RNA polymerase are really necessary or even present. While it has been shown that mutations in site I completely abolish CcpN-mediated regulation,[9] this has not been proven for site II. It has been demonstrated that the binding efficiency to two binding sites is orders of magnitude larger than for single sites,[12] so it is conceivable that one of the sites is only an auxiliary site whose sole purpose is to increase the affinity for CcpN at this promoter. These hypotheses will be tested in future investigations.

3.5. Materials and Methods

Enzymes and chemicals

Chemicals used were of the highest purity available. *E. coli* RNA polymerase and all chemicals were purchased from Sigma-Aldrich™. Taq-polymerase for cloning was purchased from Roche (Germany) and Taq-polymerase for sythesis of *in vitro* transcription templates was purchased from Solis Biodyne (Estonia).

Strains, media and growth conditions

B. subtilis strain DB104 (*ccpA::cat*)[9] was used for the preparation of *B. subtilis* RNA polymerase. *B. subtilis* strains DB104, QB5331 and SU22 were used for β-galactosidase measurements. The genotypes of these strains can be found in Table 3. TY medium (16 g Bacto tryptone, 10 g Yeast extract, 5 g NaCl in 1 l) was used as a complex medium for the purification of RNAP. SP medium (8 g nutrient broth, 0.25 g $MgSO_4$ x 7 H_2O, 1 g KCl, 1 ml $CaCl_2$ (0,5 M), 1 ml $MnCl_2$ (10 mM), 2 ml ammonium iron citrate (2,2 mg/ml)) was used as a glucose-free medium for β-galactosidase measurements.

Table 3: Strains and plasmids used in this study

Strain	relevant Genotype	Reference
E. coli TG1	wild type	Gibson (1984)[41]
B. subtilis *DB104*	wild type	Kawamura et al., (1984)[42]
B. subtilis QB5331	trpC2 levR8 grl-18 amyE::(bglP-lacZ phl)	Stülke et al. (2001)[43]
B. subtilis SU22	gapA':: pMUTIN2	Fillinger et al. (2000)[5]

Plasmid	Description	Reference
pAC6	pBR322 based vector for integration of transcriptional lacZ fusions into *amyE* locus of *B. subtilis*, Ap^R, Cm^R	Stülke et al. (1997)[44]
pACT87	pAC6 with p_{SR1} and 87 bp upstream of -35 box	Licht et al. (2005)[9]
pV6HK127A	pV6H carrying CcpN with a K127A mutation	Chaix et al., manuscript in preparation
pOU75	pOU71 based vector for IPTG-inducible expression of *ccpN* in trans, $Phleo^R$	this work
pGKI	pGK13 based plasmid carrying the *lac* repressor gene, Em^R	Brantl, unpublished
pRS6	pBR322 based plasmid carrying *lacI*, Ap^R	Brantl et al. (1992)[45]

| pPS4 | pUC19 based plasmid carrying p$_{Spac}$, Ap^R | Brantl et al. (1992)[45] |

Construction of plasmids for transcriptional *lacZ* fusions

Plasmid pAC6 (Table 3) was used to insert an EcoRI-BamHI fragment obtained by PCR from chromosomal DNA of *B. subtilis* DB104 with oligodeoxyribonucleotides SB827 (Table 4) and SB831 to obtain a transcriptional fusion of the *srl* promoter carrying 87 bp upstream of the -35 box and the promoterless *lacZ* gene. The resulting plasmid pACT87 was integrated into the *amyE* locus of strains DB104, QB5331, GP510 and SU22 and double crossing over was confirmed by streaking the chloramphenicol resistant transformants on agar with 0.5% starch and subsequent overlay with iodine/potassium-iodide solution.

Construction of a plasmid for inducible *ccpN* expression

The 2,3 kb BamHI/EcoRI fragment of plasmid pPR1 containing the *repR* gene was inserted into the pOU71 BamHI/EcoRI vector, yielding plasmid pOUR. Using primers SB 445 and SB 446 and plasmid pPR1 as template a fragment carrying the phleomycine resistance cassette flanked by EcoRI sites was generated by PCR. This fragments was cloned into the pUC19 EcoRI vector and the sequence was confirmed. The resulting vector was designated pUCP. The EcoRI fragment carrying the phleomycine resistance cassette was then obtained by restriction with EcoRI and inserted into the pOUR EcoRI vector, resulting in plasmid pOURP. Oligonucleotides SB 766 and SB 767 were annealed, yielding a polylinker sequence with 5' BamHI and 3' KpnI sticky ends. This fragment was inserted into the pOURP BamHI/KpnI vector, resulting in plasmid pOU72. A *lac* repressor fragment was obtained by restriction of plasmid pRS6 with BamHI and XbaI and inserted into the pOU72 BamHI/XbaI vector, resulting in plasmid pOU73. A p$_{Spac}$ fragment was obtained by PCR, using primers SB 768 and SB 769 with plasmid pPS4 as a template. This fragment was subsequently inserted into the pUC19 XbaI/HindIII vector and the sequence was confirmed. Thereafter, the fragment was obtained by restriction with XbaI and NcoI and inserted into the pOU73 XbaI/NcoI vector, yielding plasmid pOU74. A NcoI/NotI fragment carrying the *ccpN* gene devoid of its own promoter was generated by PCR, using primers SB 770 and SB 771 and chromosomal DNA of *B. subtilis* as a template. This fragment was cloned into the pOU74 NcoI/NotI vector, resulting in plasmid pOU75. The sequence was confirmed. Plasmid pOU75 was then used for inducible expression of *ccpN*.

Overexpression and purification of CcpN

CcpN overexpression and purification with a Ni^{2+}-NTA-agarose column was performed as published before.[9] Further purification was performed by streptomycin phosphate precipitation and dialysis against 1x TBE buffer, followed by an anion exchange chromatography on a HiLoad Q-Sepharose 16/10 column. The protein was dissolved and dialysed against 45 mM Tris/borate buffer, pH 8.3 and applied to the anion exchange column. Elution of CcpN-His$_5$ was achieved at about 200 mM NaCl by using a linear elution gradient (2 column volumes) of the same buffer containing 1 M

NaCl. The purity and activity of CcpN-His$_5$ were verified by SDS-polyacrylamide gel electrophoresis (10 %) and EMSA.

Table 4: Oligonucleotides used in this study

Name	Sequence	Purpose
SB 827	5' ACG GAA TTC TGT ATG AAG AAG ATA TTG T	construction of *lacZ* fusion
SB 831	5' GCG GGA TCC TTT CTT TTG TTG TTA TTA	
SB 422	5' TCG AGG ATC CAT GAA AGT TCA AGA AAA CGT	template for *in vitro* transcription, *sr1*
SB 342	5' CCC AGG AGA AAT TAT TAC AG	
SB 1027	5' GAG GGC AGT CAG TGC GGA GC	template for *in vitro* transcription, *gapB*
SB 1028	5' CAA TAA AAA ATA AAA AGC ATG CGG CTT TAA GCC GCA TGC TTT TTT AGC CAC AAC CTC TTT GTC GT	
SB 1029	5' AGA GTA TCC GCT CAA TGA AA	template for *in vitro* transcription, *pckA*
SB 1030	5' CAA TAA AAA ATA AAA AGC ATG CGG CTT TAA GCC GCA TGC TTT TTT TGT TGT CGC GCG AAC AGC AC	
SB 3	5' GAA ATT AAT ACG ACT CAC TAT AGG AAA CAA CGA ACT GAA TAA	template for *in vitro* transcription, *rnaII/III*
SB 4	5' GAT ATA ATG GGT TTA CAG ATA TT	
SB 445	5' GTG AAT TCG GCC ATT TTG CGT AAT AAG A	construction of pOU75
SB 446	5' GTG AAT TCG TCG ACT GAA CAG ATT AAT AAT AGA	
SB 766	5' CCC CGC GGC CGC CCC GAG CTC CCC CCA TGG CCC TCT AGA CCC G	
SB 767	5' GAT CCG GGT CTA GAG GGC CAT GGG GGG AGC TCG GGG CGG CCG CGG GGG TAC	
SB 768	5' GCG TCT AGA CTA ACA GCA CAA GAG CGG AAA	
SB 769	5' GCG AAG CTT CCA TGG GAA TTC TTA ATT GTT ATC CGC TCA CAA	
SB 770	5' GCG CCA TGG ATG AAG TGA AAA GGT GGT GAG	
SB 771	5' GCG GCG GCC GCT TAT TAT AGG ATT TCA TTT TCA GA	

Protein concentration determination

The protein concentration of CcpN-His$_5$ was determined by absorption spectroscopy using a molar extinction coefficient of 5680 l/(mol^{-1}cm^{-1}) at 280 nm according to the method of Gill and von Hippel.[22]

Circular dichroism spectroscopy

Far UV circular dichroism measurements were performed on a CD spectropolarimeter J-820 (Jasco). Spectra were recorded at a scan speed of 100 nm/min, at a response time of 2 s and accumulated. The optical path length was 1 mm and the temperature set at 20 °C. The protein concentration was 108 µg/ml (4.4 µM) in 45 mM Tris/borate buffer, pH 8.3. The effect of the metabolites on the secondary structure of CcpN-His$_5$ was determined by titration of the respective chemical compounds to the protein solution. Spectra were corrected for buffer baseline containing the respective metabolite concentration.

Preparation of templates for *in vitro* transcription

Double-stranded templates for *in vitro* transcription were obtained using the corresponding primers (Table 3) in a PCR on chromosomal DNA of *B. subtilis* DB104. The PCR products were phenolised, extracted with chloroform twice and ethanol-precipitated using 15 mg/l glycogen as carrier. Pellets were washed with 80 % EtOH and dissolved in aqua bidest. The preparation was analysed on an agarose gel and, subsequently, the DNA concentration adjusted to 1 µM.

Preparation of *B. subtilis* RNA polymerase

B. subtilis ccpN knockout strain DB104 (*ccpN::cat*)[9] was grown in TY medium to an OD$_{560}$ of 4. Cells were then harvested by centrifugation and sonicated 10 minutes in sonication buffer (40 mM potassium phosphate, 10 mM EDTA, 30 mM NaCl, 10 mM β-mercaptoethanol, 10 mM EGTA). The supernatant obtained by centrifugation was filtrated through a 100,000 Da MW cutoff Millipore column for 20 min at 6000 g to remove smaller proteins and exchange the sonication buffer for RNAP storage buffer (25 mM Tris/HCl pH 8.4, 1 mM EDTA, 7 mM β-mercaptoethanol, 50 % glycerol). The preparation was stored at -20 °C.

***In vitro* transcription**

In vitro transcription reactions were performed in a final volume of 10 µl in *in vitro* transcription buffer (40 mM Tris/acetate pH 7.5, 10 mM magnesium acetate, 100 mM potassium acetate, 20 % glycerol) in the presence of 0.1 mM ATP, CTP and GTP, 0.01 mM UTP and 0.011 µM [α-^{32}P]UTP. If indicated, potential ligands were added, then 100 nM of double stranded DNA template and 100 nM of RNA polymerase. The reaction was gently mixed and incubated for 15 min at 37 °C. One volume of formamide loading dye was added to the reaction, followed by denaturation for 5 min at 90 °C, quick cooling on ice and analysis on a 6 % denaturing polyacrylamide gel. Electrophoresis was performed at 300 V/25 mA for 50 min. Gels were dried and subjected to PhosphoImaging (Fujix BAS 1000). PC BAS 2.0e software was used for quantification of the bands.

EMSAs

Binding reactions were performed in a final volume of 10 µl in *in vitro* transcription buffer (see above), 0.05 g/l herring sperm DNA as non-specific competitor, 1 nM of end-labelled DNA fragment and 156 nM to 1.25 µM of CcpN-His$_5$. All CcpN-His$_5$ dilutions were made in storage buffer and the same volume of diluted protein was used in each sample to ensure an equal salt concentration. After incubation at 37 °C for 15 min, the reaction mixtures were separated on 8 % native polyacrylamide gels run at room temperature for 1 hour at 230 V. Gels were dried and subjected to PhosphoImaging (Fujix BAS 1000).

Acknowledgements

We thank N. Declerck and S. Aymerich for sending us the plasmid for the purification of the CcpN K127A mutant and for inspiring discussion. Furthermore, we thank Jörg Stülke for providing us with the glycolysis mutant strains. In addition, we thank E. Birch-Hirschfeld (Institut für Virologie, Jena) for synthesising the oligodeoxyribonucleotides and Nadine Möbius for helping with the construction of the p$_{SR1}$-*lacZ* fusions. Plasmid pOU71 was a kind gift from Kenn Gerdes, and we thank Sven Preis for constructing plasmid pGKI. This work was supported by grant BR1552/6-2 from Deutsche Forschungsgemeinschaft (to S. B.). A. L. is financed by a scholarship from the "Fonds der chemischen Industrie" and a scholarship from the federal state of Thuringia.

3.6. References

1. Steinmetz, M. (1993). Carbohydrate catabolism: pathways, enzymes, genetic regulation, and evolution. In *Bacillus subtilis and other gram-positive bacteria: biochemistry, physiology, and molecular genetics* (Sonenshein, A. L., Hoch, J. A. & Losick, R., eds), pp. 157-170, Am. Soc. Microbiol., Washington, DC.

2. Reizer, J., Bachem, S., Reizer, A., Arnaud, M., Saier, M. H. Jr. & Stülke, J. (1999). Novel phosphotransferase system genes revealed by genome analysis: the complete complement of PTS proteins encoded within the genome of *Bacillus subtilis*. *Microbiology* **145**, 3419-3429.

3. Chambliss, G. H. (1993). Carbon source mediated catabolite repression. In *Bacillus subtilis and other gram-positive bacteria: biochemistry, physiology, and molecular genetics* (Sonenshein, A. L., Hoch, J. A. & Losick, R., eds), pp. 212-219, Am. Soc. Microbiol., Washington, DC.

4. Stülke, J. & Hillen, W. (2000). Regulation of carbon catabolism in Bacillus species. *Annu. Rev. Microbiol.* **54**, 849-880.

5. Fillinger, S., Boschi-Muller, S., Azza, S., Dervyn, E., Branlant, G. & Aymerich, S. (2000). Two glyceraldehyde-3-phosphate dehydrogenases with opposite physiological roles in a nonphotosynthetic bacterium. *J. Biol. Chem.* **275**, 14031-14037.

6. Yoshida, K., Kobayashi, K., Miwa, Y., Kang, C. M., Matsunaga, M., Yamaguchi, H., Tojo, S., Yamamoto, M., Nishi, R., Ogasawara, N., Nakayama, T. & Fujita, Y. (2001). Combined transcriptome and proteome analysis as a powerful approach to study genes under glucose repression in *Bacillus subtilis*. *Nucleic Acids Res.* **29**, 683-692.

7. Servant, P., Le Coq, D. & Aymerich, S. (2005). CcpN (YqzB), a novel regulator for CcpA-independent catabolite repression of *Bacillus subtilis* gluconeogenic genes. *Mol. Microbiol.* **55**, 1435-1451.

8. Goldie, H. & Medina, V. (1990). Physical and genetic analysis of the phosphoenolpyruvate carboxykinase (*pckA*) locus from *Escherichia coli* K12. *Mol. Gen. Genet.* **220**, 191-196.

9. Licht, A., Preis, S. & Brantl, S. (2005). Implication of CcpN in the regulation of a novel untranslated RNA (SR1) in *Bacillus subtilis*. *Mol. Microbiol.* **58**, 189-206.

10. Heidrich, N., Chinali, A., Gerth, U. & Brantl, S. (2006). The small untranslated RNA SR1 from the *Bacillus subtilis* genome is involved in the regulation of arginine catabolism. *Mol. Microbiol.* **62**, 520-536.

11. Heidrich, N., Moll, I. & Brantl, S. (2007). *In vitro* analysis of the interaction between the small RNA SR1 and its primary target *ahrC* mRNA. *Nucleic Acids Res.* **35**, 4331-4346.

12. Licht, A. & Brantl, S. (2006). Transcriptional repressor CcpN from *Bacillus subtilis* compensates asymmetric contact distribution by cooperative binding. *J. Mol. Biol.* **364**, 434-448.

13. Fujita, M. & Sadaie, Y. (1998). Rapid isolation of RNA polymerase from sporulating cells of *Bacillus subtilis*. *Gene* **221**, 185-190.

14. Sogo, J. M., Inciarte, M. R., Corral, J., Viñuela, E. & Salas, M. (1979). RNA polymerase binding sites and transcription map of the DNA of *Bacillus subtilis* phage phi29. *J. Mol. Biol.* **127**, 411-436.

15. Brantl, S., Nuez, B. & Behnke, D. (1992). *In vitro* and *in vivo* analysis of transcription within the replication region of plasmid pIP501. *Mol. Gen. Genet.* **234**, 105-112.

16. Stülke, J., Martin-Verstraete, I., Glaser, P. & Rapoport, G. (2001) Characterization of glucose-repression-resistant mutants of *Bacillus subtilis*: identification of the *glcR* gene. *Arch. Microbiol.* **175**, 441-449

17. Soga, T., Ueno, Y., Naraoka, H., Ohashi, Y., Tomita, M. & Nishioka, T. (2002). Simultaneous determination of anionic intermediates for *Bacillus subtilis* metabolic pathways by capillary electrophoresis electrospray ionization mass spectrometry. *Anal. Chem.* **74**, 2233-2239.

18. Soga, T., Ohashi, Y., Ueno, Y., Naraoka, H., Tomita, M. & Nishioka, T. (2003). Quantitative metabolome analysis using capillary electrophoresis mass spectrometry. *J. Proteome Res.* **2**, 488-494.

19. Scott, J. W., Hawley, S. A., Green, K. A., Anis, M., Stewart, G., Scullion, G. A., Norman, D. G. & Hardie, D. G. (2004). CBS domains form energy-sensing modules whose binding of adenosine ligands is disrupted by disease mutations. *J. Clin. Invest.* **113**, 274-284.

20. Kery, V., Poneleit, L., & Kraus, J. P. (1998). Trypsin cleavage of human cystathionine beta-synthase into an evolutionarily conserved active core: structural and functional consequences. *Arch. Biochem. Biophys.* **355**, 222-232.

21. Schmidt-Rose, T. & Jentsch, T. J. (1997). Reconstitution of functional voltage-gated chloride channels from complementary fragments of Clc-1. *J. Biol. Chem.* **272**, 20515-20521.

22. Gill, S. C. & von Hippel, P. H. (1989). Calculation of protein extinction coefficients from amino acid sequence data. *Anal. Biochem.* **182**, 319-326.

23. Härtig, E., Geng, H., Hartmann, A., Hubacek, A., Münch, R., Ye, R. W., Jahn, D. & Nakano, M. M. (2004). *Bacillus subtilis* ResD induces expression of the potential regulatory genes *yclJK* upon oxygen limitation. *J. Bacteriol.* **186**, 6477–6484.

24. Tucker, N. P., D'Autréaux, B., Spiro, S. & Dixon, R. (2006). Mechanism of transcriptional regulation by the *Escherichia coli* nitric oxide sensor NorR. *Biochem. Soc. Trans.* **34**, 191-194.

25. Barragá, M. J. L., Blázquez, B., Zamarro, M. T., Mancheño, J. M., García, J. L., Díaz E., & Carmona, M. (2005). BzdR, a repressor that controls the anaerobic catabolism of benzoate in *Azoarcus sp.* CIB, is the first member of a new subfamily of transcriptional regulators. *J. Biol. Chem.* **280**, 10683-10694.

26. Laloknam, S., Tanaka, K., Buaboocha, T., Waditee, R., Incharoensakdi, A., Hibino, T., Tanaka, Y. & Takabe T. (2006). Halotolerant cyanobacterium *Aphanothece halophytica* contains a betaine transporter active at alkaline pH and high salinity. *Appl. Environ. Microbiol.* **72**, 6018-6026.

27. St. Leger, R. J., Joshi, L. & Roberts, D. (1998). Ambient pH is a major determinant in the expression of cuticle-degrading enzymes and hydrophobin by *Metarhizium anisopliae*. *Appl. Environ. Microbiol.* **64**, 709-713.

28. Fauquant, C., Diederix, R. E., Rodrigue, A., Dian, C., Kapp, U., Terradot, L., Mandrand-Berthelot, M. A. & Michaud-Soret, I. (2006). pH dependent Ni(II) binding and aggregation of *Escherichia coli* and *Helicobacter pylori* NikR. *Biochimie* **88**, 1693-1705.

29. Nakano, M. M. & Hulett, F.M. (1997). Adaptation of *Bacillus subtilis* to oxygen limitation. *FEMS Microbiol. Lett.* **157**, 1–7.

30. Tobisch, S., Zühlke, D., Bernhardt, J., Stülke, J. & Hecker, M. (1999). Role of CcpA in regulation of the central pathways of carbon catabolism in *Bacillus subtilis*. *J. Bacteriol.* **181**, 6996–7004.

31. Schumacher, M. A., Miller, M. C. & Brennan, R. G. (2004) Structural mechanism of the simultaneous binding of two drugs to a multidrug-binding protein. *EMBO J.* **23**, 2923-2930.

32. Ramsay, R. R., Jones, T. Z. & Hynson, R. M. (2005). Alteration in spectral properties on ligand binding reveals flexibility in monoamine oxidase. *Med. Sci. Monit.* **11**, 15-20.

33. Voelker, U., Voelker, A., Maul, B., Hecker, M., Dufour, A. & Haldenwang, W. G. (1995). Separate mechanisms activate sigma B of *Bacillus subtilis* in response to environmental and metabolic stresses. *J. Bacteriol.* **177**, 3771-3780.

34. Maul, B., Völker, U., Riethdorf, S., Engelmann, S. & Hecker, M. (1995) Sigma B-dependent regulation of *gsiB* in response to multiple stimuli in *Bacillus subtilis*. *Mol. Gen. Genet.* **248**, 114-120.

35. Stevens, R. C. & Lipscomb, W. N. (1990). Allosteric control of quaternary states in *E. coli* aspartate transcarbamylase. *Biochem. Biophys. Res. Commun.* **171**, 1312-1318.

36. Picossi, S., Belitsky, B. R. & Sonenshein, A. L. (2007) Molecular mechanism of the regulation of *Bacillus subtilis gltAB* expression by GltC. *J. Mol. Biol.* **365**, 1298-1313.

37. Commichau, F. M., Herzberg, C., Tripal, P., Valerius, O. & Stülke, J. (2007). A regulatory protein-protein interaction governs glutamate biosynthesis in *Bacillus subtilis*: the glutamate dehydrogenase RocG moonlights in controlling the transcription factor GltC. *Mol. Microbiol.* **65**, 642-654.

38. Escolar, L., Perez-Martin, J. & de Lorenzo, V. (1998). Coordinated repression *in vitro* of the divergent *fepA-fes* promoters of *Escherichia coli* by the iron uptake regulation (Fur) protein. *J. Bacteriol.* **180**, 2579–2582.

39. Heltzel, A., Lee, I. W., Totis, P. A. & Summers, A. O. (1990). Activator-dependent preinduction binding of sigma-70 RNA polymerase at the metal-regulated *mer* promoter. *Biochemistry* **29**, 9572–9584.

40. Monsalve, M., Mencia, M., Salas, M. & Rojo, F. (1996). Protein p4 represses phage phi 29 A2c promoter by interacting with the alpha subunit of *Bacillus subtilis* RNA polymerase. *Proc. Natl. Acad. Sci. USA* **93**, 8913–8918.

41. Baer, R., Bankier, A. T., Biggin, M. D., Deininger, P. L., Farrell, P. J., Gibson, T. J., Hatfull, G., Hudson, G. S., Satchwell, S. C., Séguin, C., Tuffnell, P. S. & Barrell, B. G. (1984). DNA sequence and expression of the B95-8 Epstein-Barr virus genome. *Nature* **310**, 207-211.

42. Kawamura, F. & Doi, R. H. (1984). Construction of a *Bacillus subtilis* double mutant deficient in extracellular alkaline and neutral proteases. *J. Bacteriol.* **160**, 442–444.

43. Stülke, J., Martin-Verstraete, I., Glaser, P. & Rapoport G. (2001). Characterization of glucose-repression-resistant mutants of *Bacillus subtilis*: identification of the *glcR* gene. *Arch. Microbiol.* **175**, 441-449.

44. Stülke, J., Martin-Verstraete, I., Zagorec, M., Rose, M., Klier, A. & Rapoport, G. (1997). Induction of the *Bacillus subtilis ptsGHI* operon by glucose is controlled by a novel antiterminator, GlcT. *Mol. Microbiol.* **25**, 65-78.

45. Brantl S. & Behnke D. (1992). The amount of RepR protein determines the copy number of plasmid pIP501 in *Bacillus subtilis*. *J. Bacteriol.* **174**, 5475-5478.

4. The transcriptional repressor CcpN from *Bacillus subtilis* uses different repression mechanisms at different promoters.

Andreas Licht* & Sabine Brantl

AG Bakteriengenetik, Friedrich-Schiller-Universität Jena, D-07743 Germany

Published in: *Journal of Biological Chemistry*, **284**: 30032-30038 (2009)

*corresponding author

4.1. Summary

CcpN, a transcriptional repressor from *Bacillus subtilis* that is responsible for the carbon catabolite repression of three genes, has been characterised in detail in the past 4 years. However, nothing is known about the actual repression mechanism so far. Here, we present a detailed study on how CcpN exerts its repression effect at its three known target promoters of the genes *sr1*, *pckA* and *gapB*. Using gel shift assays under non-repressive and repressive conditions, we showed that CcpN and RNA polymerase can bind simultaneously and that CcpN does not prevent RNA polymerase (RNAP) binding to the promoter. Furthermore, we investigated the effect of CcpN on open complex formation and demonstrate that CcpN also does not act at this step of transcription initiation. Investigation of abortive transcript synthesis revealed that CcpN acts differently at the three promoters: At the *sr1* and *pckA* promoter, promoter clearance is impeded by CcpN while synthesis of abortive transcripts is repressed at the *gapB* promoter. Eventually, we demonstrated with far western blots and co-elution experiments that CcpN is able to interact with the RNAP α-subunit, which completes the picture of the requirements for the repressive action of CcpN. On the basis of the presented results we propose a new working model for CcpN action.

4.2. Introduction

CcpN, a transcriptional repressor from *B. subtilis*, is mediating CcpA-independent carbon catabolite repression of at least three genes: *sr1*, encoding a small RNA and *pckA* and *gapB* (1,2), encoding two gluconeogenic enzymes (3,4). Since its discovery in 2005, CcpN has been thoroughly investigated: Binding properties and binding motives were examined, revealing that CcpN possesses two asymmetric binding sites which are bound cooperatively and positioned differently at the three regulated promoters (5): At the *sr1* promoter, binding sites are located upstream of the -35 region and between the -35 and the -10 region, while binding sites cover the -35 as well as the -10 region at the *pckA* promoter. One operator at the *gapB* promoter overlaps the -10 region, the second one is located around +20. ATP and low pH have been identified as signals required for CcpN-mediated repression (6) and the detailed biophysical properties of CcpN-DNA interaction have been reported (7). In addition, it has been shown that CcpN is controlling central carbon fluxes in the metabolism of *B. subtilis* and that the growth defect of CcpN knockout mutants is caused by ATP dissipation via extensive futile cycling (8). It has been demonstrated that a CcpN knockout is able to increase the industrial production of riboflavin in *B. subtilis* by a deregulation of the *gapB* gene (9). However, nothing is known about the actual repression mechanism of CcpN yet.

Initiation of transcription is a stepwise process (10), beginning with binding of RNA polymerase to the promoter and formation of a loose closed complex, which is then rearranged into a tighter closed complex. This is followed by the melting of DNA around the transcriptional start site, called the open complex. RNAP can subsequently form the initiation complex and begin to transcribe the DNA, often producing short abortive transcripts resulting from failed attempts to leave the promoter. Eventually, RNAP escapes the promoter and forms the elongation complex. Transcriptional repressors can act at any of these steps, beginning with steric hindrance of RNAP binding, like the Fur protein of *Escherichia coli* (11) over the inhibition of open complex formation, like *B. subtilis* SpoOA at the *abrB* promoter (12) to the prevention of promoter clearance, as observed with the phage Φ29 protein p4 at the viral A2c promoter (13). Different mechanisms of transcriptional repression have already been reviewed in detail (14).

While steric hindrance of RNA polymerase binding does not involve direct repressor-RNAP contacts, repression of other steps in the transcription initiation process often does. In most of those cases, contacts between a transcriptional repressor and the C-terminal domain of the α-subunit of RNAP are described, as for the p4 protein at the A2c promoter or for the repressor Spx from *B. subtilis* (13,15). However, interactions with other subunits of RNAP have also been proposed, for example for the Rsd protein of *E. coli* or the main carbon catabolite mediator of *B. subtilis*, CcpA (16,17). A special case of repressors that interact with RNA polymerase subunits are anti-σ-factors. These proteins can sequester free σ-factor and are thus able to influence the expression of whole regulons (18,19).

In this work, we present a detailed analysis of the action of CcpN at all steps of transcription initiation and show that it prevents promoter clearance at the *srl* and *pckA* promoter, while displaying a rare effect at the *gapB* promoter: It allows the formation of the open complex, but prevents the synthesis of abortive transcripts. Furthermore, we demonstrate that CcpN is able to interact with the α-subunit of RNAP and probably regulates the *srl* and *pckA* promoters this way. Eventually, we present a new working model for CcpN-mediated transcriptional repression in regard to the specific operator positions and promoter sequences.

4.3. Results

CcpN does not inhibit formation of the closed complex.

Transcriptional repressors can act during a variety of different steps in transcription initiation. To investigate whether CcpN exerts its repression effect by preventing RNA polymerase binding to the promoter, we performed gel shift assays using 89 bp end-labelled double-stranded DNA fragments carrying the *srl*, *pckA*, *gapB* or RNAIII (as a negative control that is unable to bind CcpN) promoters (Figure 1). Purified CcpN-His$_5$ and purified His-tagged *B. subtilis* RNA polymerase alone and together were incubated with the labelled DNA fragment and complex formation was analysed on native polyacrylamide gels. The presence of CcpN or RNA polymerase alone resulted in a single band corresponding to the respective protein-DNA-complex at all three promoters. When both proteins were present, an additional band was visible at all three promoters, emerging from a complex of DNA, CcpN and RNA polymerase. As expected, the control promoter of RNAIII showed only a single band corresponding to an RNAP-DNA complex, but no CcpN-DNA-complex. All experiments were performed under non-repressive (0 mM ATP, pH 7.3) and under repressive conditions (3 mM ATP, pH 6.5) to assay if CcpN is able to prevent RNA polymerase binding to the promoter sequence. For analysis under repressive conditions, both ATP and low pH were also present in the gel and in the running buffer to ensure that the conditions did not change during electrophoresis. At all three promoters, the intensity of the band representing the CcpN-RNAP-DNA complex did not change in intensity when comparing non-repressive with repressive conditions, indicating that CcpN is not able to prevent the formation of the closed complex.

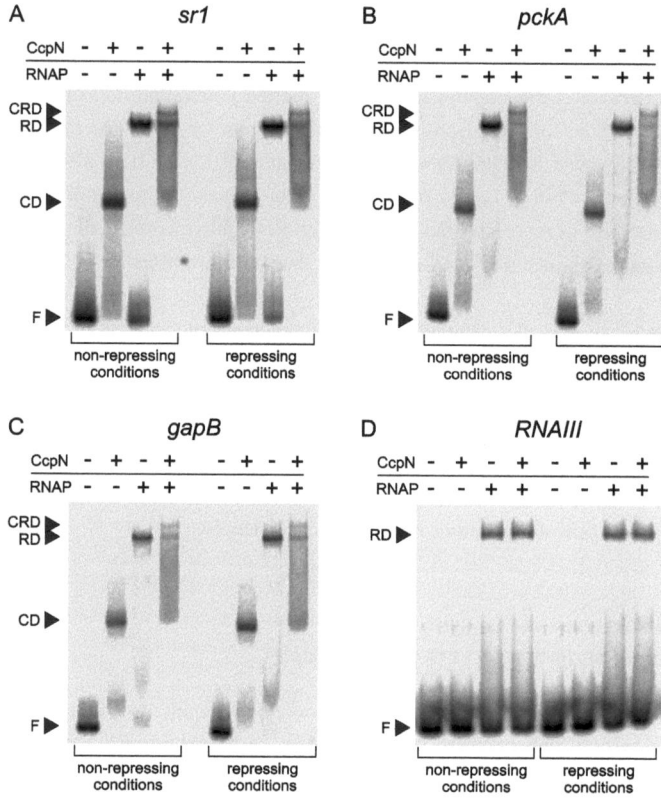

Figure 1: EMSAs with His-tagged CcpN and RNA polymerase at the *sr1* (A), *pckA* (B), *gapB* (C) and *RNAIII* (D) promoters.

The presence or absence of 3 μM CcpN-His$_5$ and 3 μM RNAP-His$_6$ is indicated above each lane. Experiments were performed under non-repressing (pH 7.0; 0 mM ATP) or repressing (pH 6.5; 3 mM ATP) conditions. Autoradiograms of the gels are shown. F: free DNA, CD: CcpN-DNA-complex, RD: RNAP-DNA-complex, CRD: CcpN-RNAP-DNA-complex.

CcpN does not inhibit open complex formation.

The next step in transcription initiation is the formation of the open complex, involving melting of the DNA at the promoter region. In order to detect formation of an open complex, a double stranded DNA fragment was probed for the presence of single-stranded regions under

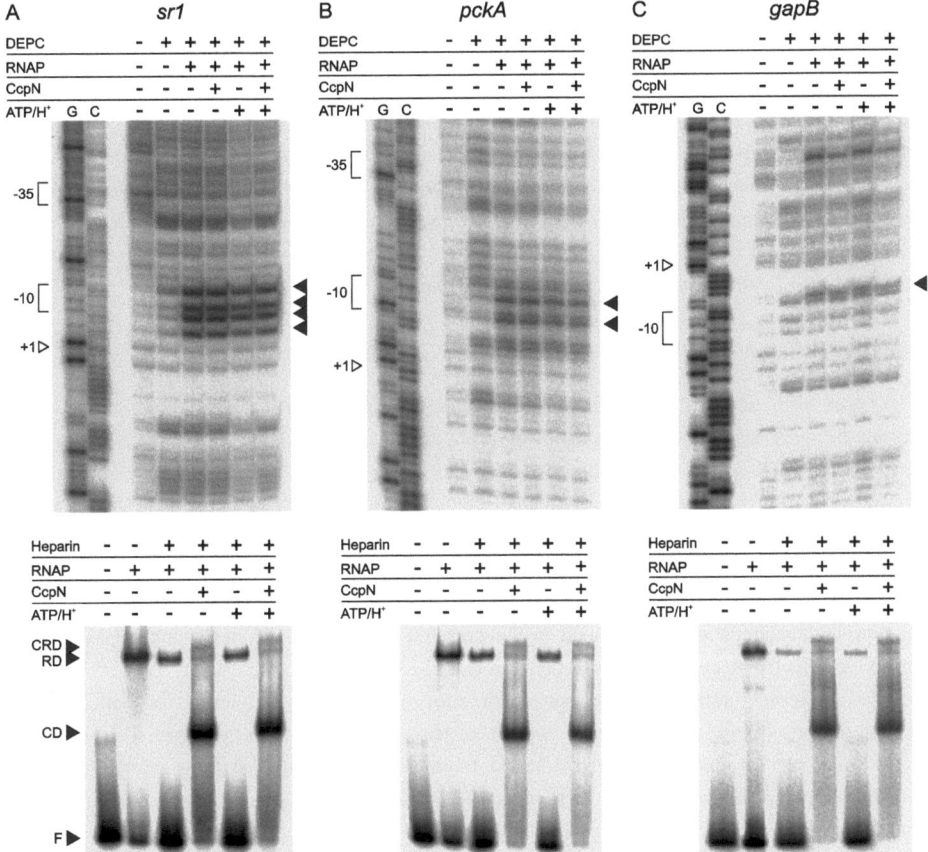

Figure 2: Open complex formation assay at the *sr1* (A), *pckA* (B) and *gapB* (C) promoters.

Probing with DEPC is shown at the top while the corresponding Heparin-probing is shown below. For DEPC-probing, DEPC (10 %), RNAP (100 nM), CcpN-His$_5$ (100 nM), were added where indicated. Bands showing the presence of single stranded DNA regions and therewith open complexes are indicated by arrows. G: G>A sequencing reaction, C: C+T sequencing reaction. Positions of +1, the -10 and -35 box are indicated. Please note that the noncoding strand was used for *sr1* and *pckA*, while the coding strand was used for *gapB*. For Heparin-probing, Heparin (0.1 g/l) CcpN-His$_5$ (3 µM), RNAP-His$_6$ (3 µM) ATP (3 mM) or HCl (to a final pH of 6.5) were added where indicated. F: free DNA, CD: CcpN-DNA-complex, RD: RNAP-DNA-complex, CRD: CcpN-RNAP-DNA-complex. Autoradiograms of the gels are shown.

different conditions using DEPC (Figure 2), which is known to react preferentially with single-stranded regions in B-form DNA (23). Usually KMnO$_4$ is used for detection of single- stranded regions, but did not work under our buffer conditions. Therefore, DEPC was used, although it has the disadvantage of producing weaker signals at stacked adenosine residues. As shown in Figure 2, signals emerged at all three promoters upon addition of RNA polymerase that were not present in the negative control, where only DEPC was added. These signals persisted in the presence of CcpN (non-repressive conditions) as well as in the presence of CcpN, ATP and low pH (repressive conditions) at all investigated promoters. Thus, one can conclude that CcpN is not able to prevent formation of the open complex at any of the three promoters. To corroborate these findings, another assay for open complex formation using Heparin as a probe has been performed. Again, CcpN was not able to prevent the formation of open complexes at repressive or non-repressive conditions at any of the investigated promoters (Figure 2, bottom panels).

CcpN acts differently at the three promoters.

Since formation of the open complex is not impeded by CcpN at any promoter, it can either prevent the synthesis of abortive transcripts or promoter clearance. To investigate this issue, *in vitro* transcription reactions under non-repressive and repressive conditions were performed and analysed on two different denaturing polyacrylamide gels: 6 % gels were used to detect full-length transcripts while 23 % gels were used to detect abortive transcripts (Figure 3). Since there is no uridine within the first 11 bases of the *sr1* transcript, [α-^{32}P]ATP instead of UTP was used for labelling. This resulted in very faint bands for both the full-length and the abortive transcripts, because all *in vitro* transcription reactions were performed in the presence of 3 mM ATP necessary to observe the repressive effect of CcpN. To ensure that the observed abortive transcripts are produced by the analysed promoters rather than non-promoter sites on the template, templates with mutated promoters were investigated (Figure 3). Indeed, certain transcripts within the expected size of 3-11 nt are no longer produced from the mutated fragments, indicating that they emerge from the investigated promoters. At all three promoters, formation of full-length or abortive transcripts was not influenced in the presence of CcpN or low pH alone. Figure 3A shows that abortive transcripts are produced at the *sr1* promoter in all four lanes, even under repressive conditions, while synthesis of the full-length transcript is significantly repressed in the presence of CcpN, ATP and low pH. At the *pckA* promoter, most of the abortive transcripts are still produced during CcpN-mediated repression, however, the smallest two transcripts are lost. Nevertheless, Figure 3B clearly shows that abortive transcription in general is not affected by CcpN. A completely different picture can be found at the *gapB* promoter (Figure 3C). Here, bands corresponding to abortive transcripts are hardly or not at all detectable under repressive conditions. Thus, one can conclude that CcpN acts at the *sr1* and *pckA* promoters by preventing RNA polymerase from leaving the promoter and proceeding with transcription, while still allowing the production of short abortive transcripts. At the *gapB* promoter, however, CcpN impedes transcription initiation itself, resulting in the inability to produce abortive transcripts.

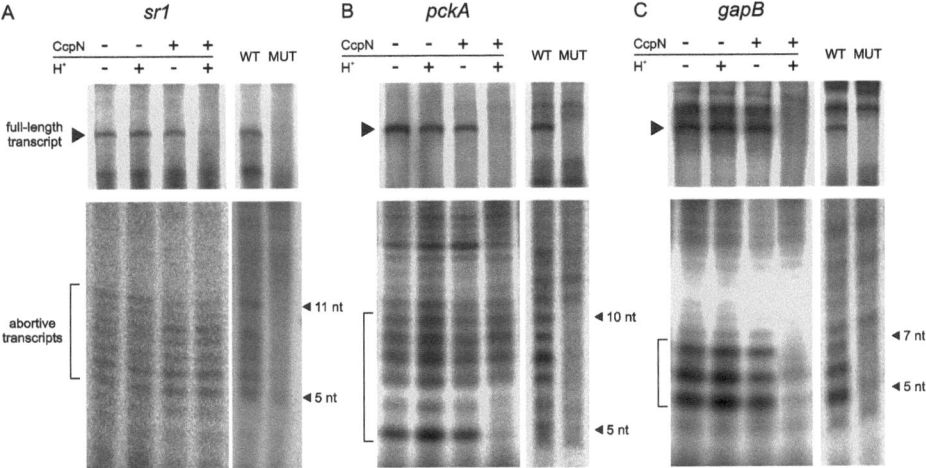

Figure 3: *In vitro* transcription and detection of abortive transcripts at the *sr1* (A), *pckA* (B) and *gapB* (C) promoters.

Transcription was performed in *in vitro* transcription buffer (see experimental procedures) using 100 nM DNA template and 100 nM His-tagged *B. subtilis* RNA polymerase. 300 nM CcpN-His$_5$ was added or pH was lowered where indicated. Half of each reaction was separated on either a 6 % denaturing polyacrylamide gel to detect the full length transcripts, indicated by an arrow or on a 23 % denaturing polyacrylamide gel to detect abortive transcripts, indicated by a bracket. Control experiments to the right of each panel show which of the abortive transcripts are produced by the investigated promoters. WT: wild-type promoter, MUT: mutated promoter, where the -10 regions have been replaced by the sequence GCCGAT (*sr1*) or GCCGCT (*pckA* and *gapB*). The estimated size of the abortive transcripts on each gel is indicated by arrows. Autoradiograms of the gels are shown.

CcpN is able to interact with RNA polymerase

Since CcpN is able to prevent RNA polymerase from leaving the promoter, we wanted to find out whether this is due to a direct interaction. To this end, we purified the RNA polymerase α-subunit (RpoA) as well as the *B. subtilis* major σ-factor SigA. Figure 4A shows the two proteins, along with BSA and purified CcpN. While SigA is apparently pure, the α-subunit contains some impurities, although in a much lower concentration than the protein itself. Two gels with identical protein samples have subsequently been subjected to far western blotting to analyse possible interactions between CcpN and these proteins (Figure 4B). The left panel shows the control blot that was only incubated with primary (anti-CcpN) and secondary antibody. As expected, CcpN itself produced a very strong signal, indicating that the antibodies work as intended. However, there are also two signals in the lane with the RpoA preparation: A very intensive signal corresponding to the largest impurity, indicating extensive antibody cross-reaction and a weak signal at the 27 kDa impurity. The RpoA band itself did not produce a signal, demonstrating that the anti-CcpN-antibody did not bind to it unspecifically. Furthermore, there were no antibody cross reactions with either SigA or BSA. The right panel shows the experiment itself, where the blot has been incubated with CcpN before the application of the first antibody. Strikingly, a band emerges that corresponds exactly to the 39 kDa band comprised of RpoA, indicating that CcpN is able to specifically interact with the RNA polymerase α-subunit. Neither SigA nor BSA showed any interaction with CcpN at all.

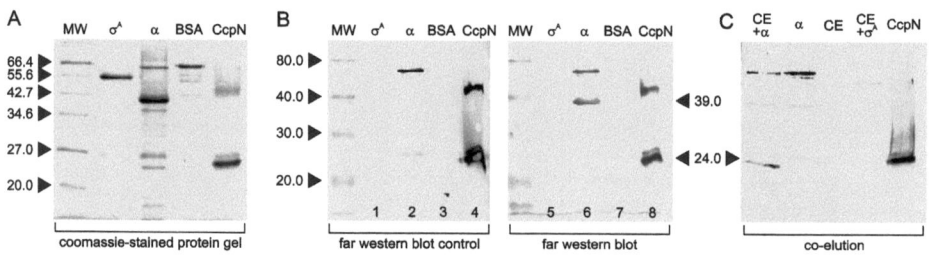

Figure 4: CcpN-RpoA interaction studies.

The corresponding molecular weights of the marker bands are indicated beside the marker lanes. (A) 15.5 % SDS-polyacrylamide gel of different purified proteins. MW: molecular weight marker; σ: purified SigA-His$_6$; α: purified RpoA-His$_6$; BSA: bovine serum albumin; CcpN: purified CcpN-His$_5$. 1 µg of each protein was loaded into each lane. (B) Far western blot of the protein gel shown in (A). Equal amounts of protein were loaded into lanes 1-4 and 5-8, respectively. Proteins were renatured after blotting by washing with SDS-free PBS. Lanes 1-4 are control lanes and were just blocked, incubated with rabbit-anti-CcpN antibody and subsequently with horseradish peroxidase coupled anti-rabbit antibody. Lanes 5-8 are the sample lanes and were treated like lanes 1-5, but were incubated with 200 nM CcpN-His$_5$ after blocking and before incubation with anti-CcpN antibody. The blots were developed using horseradish peroxidase catalysed conversion of diaminobenzidine. PC-BAS 2.08e software was used for quantification. (C) Co-elution of RpoA-His$_6$ and CcpN. The lanes were loaded as follows: CE+α: RpoA-His$_6$ preincubated with *B. subtilis* DB104 protein crude extract (see experimental procedures) and subsequently purified using a Ni^{2+}-NTA-agarose column; α: RpoA-His$_6$ without preincubation with *B. subtilis* DB104 protein crude extract; CE: *B. subtilis* DB104 protein crude extract, purified; CE+σ: SigA-His$_6$ preincubated with *B. subtilis* DB104 protein crude extract and subsequently purified; CcpN: purified CcpN-His$_5$. Equal amounts of eluate were loaded into each lane.

To corroborate these findings, we investigated whether CcpN can be co-eluted with an α subunit preparation. To this end, a crude extract of *B. subtilis* DB104 was incubated with RpoA-His$_6$ and subsequently purified using a Ni^{2+}-NTA-agarose column. As controls, RpoA-His$_6$ alone, the crude extract alone as well a crude extract preincubated with SigA-His$_6$ were purified in the same manner. Figure 4C shows the results of these experiments. It can be clearly seen that only in the case where the crude extract was preincubated with RpoA, a new band emerges that corresponds to native CcpN. As expected, this band runs marginally faster than the purified CcpN due to the lack of the His-Tag used for CcpN purification. Taken these results and the far western blot together, one can conclude that CcpN is able to specifically interact with the α-subunit of RNA polymerase.

4.4. Discussion

Repression mechanism of CcpN

Here, we report the elucidation of the repression mechanism employed by the transcriptional repressor CcpN from *B. subtilis*. Gel shift assays demonstrated that CcpN does not prevent RNA polymerase binding and that both proteins can bind simultaneously to the promoter. Interestingly, CcpN and RNA polymerase, although able to bind simultaneously, appear to compete for binding to the used DNA fragments. Figure 1 clearly shows that the bands for all three complexes are significantly weaker when both proteins are present than the complexes where only one of the proteins is present. Since CcpN and RNA polymerase concentrations have been chosen to reflect their actual concentrations *in vivo* (unpublished observation, 24), it is conceivable that there is also a competition between these two proteins for promoter binding within the cell. This finding would also explain the observations made by Servant *et al.*, who reported a significant derepression of the *pckA* and *gapB* promoters in a *ccpN* knockout mutant, even under gluconeogenic conditions where CcpN is not active (3), a feature that was also reported for other transcriptional repressors, although not to such a huge extent (25).

Repressors that bind simultaneously with RNA polymerase, either at overlapping or at different sites, often repress transcription by preventing melting of DNA at the transcriptional start site, i.e. formation of the open complex. Such transcription factors are for example *E. coli* MerR at the *merT* promoter (26,27), which binds together with RNAP at opposite sites of the DNA helix, or the KorB protein of broad host range plasmid RK2 (28), whose binding sites do not overlap those of RNAP. CcpN features both versions of operator sites, some overlap with RNAP binding sites whereas some do not (5). However, open complex formation assays clearly ruled out the possibility that CcpN acts by preventing DNA melting at any of the promoters.

The inhibition of the synthesis of abortive transcripts, as observed by us at the *gapB* promoter, is a case rarely reported in literature. The H-NS protein at the *rrnB* P1 promoter or the FIS protein at the *gyrB* promoter are two examples for this kind of repression (29,30). For H-NS, a binding pattern similar to CcpN has been reported, where the operator overlaps the RNAP binding site. H-NS is then able to alter the DNA structure at this position, allowing the formation of open complexes, but preventing subsequent transcription. A similar mode of action is conceivable for CcpN at the *gapB* promoter. DNase I footprints have revealed the appearance of several hypersensitive sites upon CcpN binding at this promoter, which is usually a good indication for structural alterations of the DNA (3,4). At the *srl* and *pckA* promoter, however, abortive transcripts are readily formed, but escape of RNA polymerase from the promoter is inhibited. Prevention of promoter clearance is usually mediated by one of two different ways: A repressor can bind downstream of RNAP and simply create a roadblock before a stable elongation complex can be formed. This has for example been shown for CcpA-mediated regulation of the *treP* gene in *B. subtilis*, and even as a prove of principle with an artificial construct using the Lac repressor (31,32). Regarding the operator positions at the *sr1* and *pckA* promoters, this mechanism appears to be highly unlikely, which favours the second alternative possibility: An interaction between the repressor molecule and parts of the RNA polymerase. It is known that the polymerase can be stalled at promoters with close-to-consensus sequences, resulting from a extremely tight binding that subsequently makes promoter clearance very difficult (33). Transcriptional repressors, which usually bind their operator sequences with high affinity, can mimic the aforementioned effect by binding RNAP and keeping it in place. Examples for this mechanism include the phage Φ29 protein p4 at the phage A2c promoter (34) and the Gal repressor (35).

CcpN interacts with the RNAP-α-subunit

With respect to our finding that CcpN is able to specifically interact with the RNA polymerase α-subunit, we conclude that CcpN acts as a repressor at the *sr1* and *pckA* promoters by keeping RNAP in place through the aforementioned interaction. There are various reports about the α-subunit, and especially the C-terminal domain, being an interaction interface for transcriptional repressors, as mentioned above. However, interactions with the α-subunit have also been reported for activators, like CcpA at the *ackA* promoter (36,37) or SoxS during oxidative stress conditions (38). Considering the binding site position of CcpN at the *sr1* and *pckA* promoters, an interaction with the α-subunit appears very conceivable. It has been shown that up elements in *B. subtilis* have a slightly broader tolerance regarding location than in *E. coli* (39,40), reaching approximately from -40 to -66, which would position the α-C-terminal domain to be able to interact with CcpN at these promoters.

At the *gapB* promoter, however, an interaction with the α-subunit can be excluded, since both operator sites are too far downstream to allow any contact between the two proteins. Two possibilities are conceivable how CcpN exerts its action here: Either CcpN alters the DNA structure as mentioned above, or it interacts with an RNAP subunit other than the α-subunit or the σ-factor, since the first one cannot be contacted and no interaction has been detected with the latter. Reports of transcription factors that interact with e.g. the β-subunit are quite uncommon. One of these examples is the AsiA protein from bacteriophage T4 (41), another being the Rsd protein of *E. coli* (42), both of which have been shown to be able to interact with the core RNA polymerase. If CcpN actually interacts with parts of the RNAP other than the α-subunit needs to be experimentally determined. However, the relatively small size of CcpN, leaving not much space for extensive interaction surfaces and the fact that DNA structure is altered upon CcpN binding seem to favour the possibility of repression by DNA-structure rearrangements.

The example of CcpN shows that one single repressor can exert repression in very different ways, depending on how its operators are positioned relative to the RNA polymerase binding sites. Varying binding site distribution is quite common, found e.g. in the case of CytR from *E. coli* (43) and many more. Interestingly, cases where variations in operator positioning result in different repression mechanisms have not been frequently reported in literature. However, this is mostly because the actual repression mechanism for these proteins has not been elucidated. A well documented example where operator site positions have an impact on the repression mechanism is cre element positioning, allowing CcpA to exert a broad range of repression or even activation mechanisms on its targets (17).

Taking all results together, a quite clear picture of the repression mechanism of CcpN can be established where CcpN and the α-subunits are in a spatial position that allows interaction and subsequent promoter arrest at the *srl* and *pckA* promoters, but not at the *gapB* promoter. Here, repression by modification of the DNA structure appears to be a probable alternative.

4.5. Experimental Procedures

Strains and media used in this study

B. subtilis strain NIG2001 was used for expression of His-tagged *B. subtilis* RNA polymerase (20) and strain DB104 (21) was used for the preparation of *B. subtilis* protein crude extracts. *E. coli* strain TG1 (pREP4, pQGDR) was used for overexpression and purification of CcpN-His$_5$ and strain BL21 (DE3) (pETSigA) was used for overexpression and purification of His-tagged *B. subtilis* SigA (4,22). All strains were grown in TY medium (16 g Bacto tryptone, 10 g Yeast extract, 5 g NaCl in 1 l) with the respective antibiotics.

Protein purification

CcpN overexpression and purification with a Ni^{2+}-NTA-agarose column and by anion exchange chromatography was performed as published before (6). Expression and purification of His-tagged *B. subtilis* RNA polymerase and His-tagged SigA with a Ni^{2+}-NTA-agarose column was carried out according to the protocols established by Fujita and Sadaie (20,22).

Gel shift assays

Binding reactions were performed in a final volume of 10 µl in either 0.5x TBE and 10 mM $MgCl_2$ for the formation of closed complexes or in *in vitro* transcription buffer (40 mM Tris/acetate, pH 7.3, 10 mM magnesium acetate, 100 mM potassium acetate and 20 % glycerol) for the formation of open complexes, 0.05 g/l herring sperm DNA as nonspecific competitor and 1 nM endlabelled DNA fragment. Where indicated, 3 µM of CcpN-His$_5$, 3 µM of RNAP-His$_6$, 3 mM ATP, HCl to a final pH of 6.5 or 0.1 g/l of Heparin were added. After incubation at 37 °C for 15 min, the reaction mixtures were denatured and separated on 5 % native polyacrylamide gels run at room temperature for 1 h at 230 V. Gels were dried and subjected to PhosphorImaging (Fujix BAS 1000).

Open complex formation assays

Binding reactions were performed in a final volume of 10 µl in 50 mM sodium-cacodylate buffer (pH 7.3) using 1 nM of an endlabelled DNA fragment. Where indicated, 100 nM of CcpN-His$_5$, 100 nM of native RNAP, 3 mM ATP and/or HCl to a final pH of 6.5 were added. After 15 min incubation at 37 °C, 1 µl of DEPC (final concentration of 10 %) was added and the reaction was incubated at 37 °C for further 10 min. The reaction was stopped by the addition of 50 µl of stop solution (1.5 M NaAc, 0.1 g/l tRNA) and precipitated with ethanol, followed by dissolving of the pellet in 10 % piperidine and cleavage at the modified sites for 30 min at 90 °C. Subsequently, the cleavage reaction was precipitated with ethanol again and the pellet dissolved in formamide loading dye to a final activity of 2000 cpm/µl. Afterwards, 3 µl were denatured and separated on a 6 % denaturing polyacrylamide gel. Gels were dried and subjected to PhosphorImaging (Fujix BAS 1000).

In vitro **transcription**

In vitro transcription reactions at the *pckA* and *gapB* promoters were performed in a final volume of 20 µl in *in vitro* transcription buffer in the presence of 3 mM ATP, 0.1 mM CTP and GTP, 0.01 mM UTP and 0.011 µM [α 32P]UTP. For the *sr1* promoter, 0.1 mM UTP and 0.011 µM [α-32P]ATP were used to allow detection of abortive transcripts. Where indicated, HCl to a final pH of 6.5 and CcpN-His$_5$ were added, followed by 100 nM double-stranded DNA template and 100 nM RNAP-His$_6$. The reaction was gently mixed and incubated for 30 min at 37 °C. Half of the reaction was ethanol-precipitated with potassium acetate to keep unincorporated [α-^{32}P]NTPs in solution and then dissolved in 10 µl distilled water. One volume of formamide loading dye was added to each half of the reaction, followed by denaturation for 5 min at 90 °C, quick cooling on ice and analysis on a 6 % denaturing polyacrylamide gel to detect full-length transcripts or on a 23 % denaturing polyacrylamide gel to detect abortive transcripts. Electrophoresis was performed at 300 V/25 mA for 50 min. Gels were dried and subjected to PhosphorImaging (Fujix BAS 1000).

Western and far western blotting

For Western blotting, samples were separated on a 15.5 % SDS-polyacrylamide gel and subsequently blotted using PVDF membrane (Carl Roth, Germany). A polyclonal antiserum from rabbit against CcpN-His$_5$ as primary antibody and horseradish-peroxidase-coupled anti-rabbit secondary antibody (Santa Cruz Biotechnology, Inc., USA) was used, both with a dilution of 1:2000. Blots were developed by diaminobenzidine reaction, digitised with a ScanPrisa 640U (Acer) scanner and analysed with TINA-PC BAS 2.08e software. For far western blotting, two identical sets of protein samples were separated by SDS-PAGE and subsequently blotted. SDS was then removed from the blot membranes by washing with PBST. After blocking, the part of the membranes containing the first set of samples was incubated with blocking buffer again for 1 h at RT, while the membrane containing the other set was incubated with 200 nM CcpN-His$_5$ in blocking buffer. Both membranes were then washed with PBST and incubated with primary and secondary antibody as described in the western blotting procedure.

Co-elution

B. subtilis DB104 was grown to an OD_{560} of 4 in 150 ml TY medium, cells were harvested, resuspended in 15 ml PBS and sonicated 3 times for 10 min. 180 µl PMSF (17 g/l in isopropanol) were added prior to sonication. After centrifugation at 4 °C, the supernatant was obtained and incubated with 300 nM of purified RpoA-His$_6$, SigA-His$_6$ or without any protein for 1 h at RT. The samples were then purified using a Ni^{2+}-NTA-agarose column and the eluates analysed on a 15.5 % PAA-SDS-gel.

Acknowledgements

We are very grateful to M. Salas for the gift of native purified *B. subtilis* RNA polymerase as well as for the strain for the overproduction of His-tagged α-subunit. Furthermore, we would like to thank M. Fujita, who kindly sent us the strain for overproduction of His-tagged σ-factor. This work was supported by grant BR1552/6-3 from Deutsche Forschungsgemeinschaft (to S. B.). A. L. was financed by a scholarship from the federal state of Thuringia and by the Deutsche Forschungsgemeinschaft.

4.6. References

1. Fillinger, S., Boschi-Muller, S., Azza, S., Dervyn, E., Branlant, G., and Aymerich, S. (2000) *J. Biol. Chem.* **275**, 14031-14037

2. Yoshida, K., Kobayashi, K., Miwa, Y., Kang, C. M., Matsunaga, M., Yamaguchi, H., Tojo, S., Yamamoto, M., Nishi, R., Ogasawara, N., Nakayama, T., and Fujita, Y. (2001) *Nucleic Acids Res.* **29**, 683-692

3. Servant, P., Le Coq, D., and Aymerich, S. (2005) *Mol. Microbiol.* **55**, 1435-1451

4. Licht, A., Preis, S., and Brantl, S. (2005) *Mol. Microbiol.* **58**, 189-206

5. Licht, A., and Brantl, S. (2006) *J. Mol. Biol.* **364**, 434-448

6. Licht, A., Golbik, R., and Brantl, S. (2008) *J. Mol. Biol.* **380**, 17-30

7. Zorrilla, S., Ortega, A., Chaix, D., Alfonso, C., Rivas, G., Aymerich, S., Lillo, M. P., Declerck, N., and Royer, C. A. (2008) *Biophys. J.* **95**, 4403-4415

8. Tännler, S., Fischer, E., Le Coq, D., Doan, T., Jamet, E., Sauer, U., and Aymerich, S. (2008) *J. Bacteriol.* **190**, 6178-6187

9. Tännler, S., Zamboni, N., Kiraly, C., Aymerich, S., and Sauer, U. (2008) *Metab. Eng.* **10**(5), 216-226

10. Record, M. T. Jr., Reznikoff, W. S., Craig, M. L., McQuade, K. L., and Schlax, P.J. (1996) Escherichia coli and Salmonella typhimurium: Cellular and Molecular Biology, American Society for Microbiology, Washington DC

11. Escolar, L., Perez-Martin, J., and de Lorenzo, V. (1998) *J. Bacteriol.* **180**, 2579–2582

12. Greene, E. A., and Spiegelman, G. B. (1996) *J. Biol. Chem.* **271**, 11455-11461
13. Monsalve, M., Mencía, M., Salas, M., and Rojo, F. (1996) *Proc. Natl. Acad. Sci. USA* **93**, 8913-8918
14. Rojo, F. (2001) *Curr. Opin. Microbiol.* **4**, 145-151
15. Zhang, Y., Nakano, S., Choi, S. Y., and Zuber, P. (2006) *J. Bacteriol.* **188**, 4300-4311
16. Ilag, L. L., Westblade, L. F., Deshayes, C., Kolb, A., Busby, S. J., and Robinson, C. V. (2004) *Structure* **12**, 269-275
17. Kim, J. H., Yang, Y. K., and Chambliss, G. H. (2005) *Mol. Microbiol.* **56**, 155-162
18. Hughes, K. T., and Methee, K. (1998) *Annu. Rev. Microbiol.* **52**, 231-286
19. Campbell, E. A., Westblade, L. F., and Darst, S. A. *Curr. Opin. Microbiol.* **11**, 121-127
20. Fujita, M., and Sadaie, Y. (1998) *Gene* **221**, 185-190
21. Kawamura, F., and Doi, R. H. (1984) *J. Bacteriol.* **160**, 442-444
22. Fujita, M., and Sadaie, Y. (1998) *J. Biochem.* **124**, 89-97
23. Scholten, P. M., and Nordheim, A. (1986) *Nucleic Acids Res.* **14**, 3981-3993
24. Wagner, R. (2000) Transcription Regulation in Prokaryotes, Oxford University Press, New York
25. Barragá, M. J. L., Blázquez, B., Zamarro, M. T., Mancheño, J. M., García, J. L., Díaz, E., and Carmona, M. (2005) *J. Biol. Chem.* **280**, 10683-10694
26. Heltzel, A., Lee, I. W., Totis, P. A., and Summers, A. O. (1990) *Biochemistry* **29**, 9572-9584
27. Summers, A. O. (1992) *J. Bacteriol.* **174**, 3097-3101
28. Williams, D. R., Motallebi-Veshareh, M., and Thomas, C. M. (1993) *Nucleic Acids Res.* **21**, 1141-1148
29. Schröder, O., and Wagner, R. (2000) *J. Mol. Biol.* **298**, 737-748
30. Schneider, R., Travers, A., Kutateladze, T., and Muskhelishvili, G. (1999) *Mol. Microbiol.* **34**, 953-964
31. Ujiie, H., Matsutani, T., Tomatsu, H., Fujihara, A., Ushida, C., Miwa, Y., Fujita, Y., Himeno, H., and Muto, A. (2009) *J. Biochem.* **145**, 59-66
32. Lopez, P. J., Guillerez, J., Sousa, R., and Dreyfus, M. (1998) *J. Mol. Biol.* **276**, 861-875
33. Ellinger, T., Behnke, D., Knaus, R., Bujard, H., and Gralla, J. D. (1994) *J. Mol. Biol.* **239**, 466-475
34. Monsalve, M., Mencia, M., Rojo, F., and Salas, M. (1996) *EMBO J.* **15**, 383-391
35. Choy, H. E., Park, S. W., Aki, T., Parrack, P., Fujita, N., Ishihama, A., and Adhya, S. (1995) *EMBO J.* **14**, 4523-4529

36. Turinsky, A. J., Grundy, F. J., Kim, J. H., Chambliss, G. H., and Henkin, T. M. (1998) *J. Bacteriol.* 180, 5961-5967
37. Kim, J. H., Yang, Y. K., and Chambliss, G. H. (2005) *Mol. Microbiol.* **56**, 155-162
38. Shah, I. M., and Wolf, R. E. Jr. (2004) *J. Mol. Biol.* **343**, 513-532
39. Gourse, R. L., Ross, W., and Gaal, T. (2000) *Mol. Microbiol.* **37**, 687-695
40. Meijer, W. J., and Salas, M. (2004) *Nucleic Acids Res.* **32**, 1166-1176
41. Severinova, E., Severinov, K., and Darst, S. A. (1998) *J. Mol. Biol.* **279**, 9-18
42. Ilag, L. L., Westblade, L. F., Deshayes, C., Kolb, A., Busby, S. J., and Robinson, C. V. (2004) *Structure* **12**, 269-275
43. Collado-Vides, J., Magasanik, B., and Gralla, J. D. (1991) *Microbiol. Rev.* **55**, 371–394

5. Search for additional targets of the transcriptional regulator CcpN from *Bacillus subtilis*.

Rita A. Eckart, Sabine Brantl & Andreas Licht*

AG Bakteriengenetik, Friedrich-Schiller-Universität Jena, D-07743 Germany

Published in: *FEMS Microbiology Letters*, **299**: 223-231 (2009)

*corresponding author

5.1. Summary

Transcriptional repressor CcpN from *B. subtilis* mediates the CcpA-independent catabolite repression of three genes, *sr1* encoding a small regulatory RNA, and two gluconeogenesis genes, *gapB* and *pckA*. The intracellular concentration of CcpN was determined to be around 4000 molecules per cell. The *B. subtilis* genome was scanned for potential new CcpN target genes, out of which three showed CcpN binding activity in their upstream region. EMSAs demonstrated that the promoter regions of two putative targets, *thyB* encoding thymidylate synthase B and *yhaM* encoding a 5'-3' exoribonuclease, bound CcpN with significant affinity. A detailed contact probing of CcpN-DNA interactions revealed an interesting new binding pattern at the *thyB* promoter, where the whole promoter appears to be contacted by CcpN. Using *lacZ*-reporter gene fusions and *in vitro* transcription assays, the *thyB* promoter was investigated for a regulatory effect of CcpN. Surprisingly, CcpN does not repress transcription at this promoter, but instead acts as an activator. Alignments of the *thyB* promoters of different Gram-positive bacteria encoding CcpN revealed CcpN consensus binding sites in a significant number of them. Our data show that a bioinformatics-based approach combined with *in vivo* and *in vitro* experiments can be used to identify new targets of transcriptional regulators.

5.2. Introduction

Catabolite repression is an important regulatory aspect in a variety of bacteria, among them *Bacillus subtilis* (Chambliss, 1993; Steinmetz, 1993). In *B. subtilis*, this process is mediated primarily by the combined action of CcpA and HPr-Ser46-P by forming a transcriptional regulator upon interaction (Chambliss, 1993; Stülke & Hillen, 2000). However, carbon catabolite repression of at least three known genes, *gapB*, *pckA* and *sr1*, is mediated by the transcriptional repressor CcpN (Licht *et al.*, 2005; Servant *et al.*, 2005). CcpN binds cooperatively to two distinct binding sites at each of these promoters and has recently been shown to require ATP and a slightly acidic pH for the exertion of its repression effect, while ADP was able to counteract the ATP-mediated repression (Licht & Brantl, 2006; Licht *et al.*, 2008). The repression mechanism of CcpN has not been elucidated so far.

Bacterial transcriptional regulators can act as pure activators, e.g. MalT, the activator of maltose metabolic genes, or PhoB, an activator controlling phosphate uptake in *Escherichia coli* (Schlegel *et al.*, 2002; Yamada *et al.*, 1989), as pure repressors repressors like the Arg or Lac repressor from *E. coli* (Maas, 1994; Lewis, 2005) or as dual regulators acting either as activator or repressor as e.g. the global regulators CcpA and CodY from *B. subtilis* (Henkin, 1996; Sonenshein, 2005).

The number of genes regulated by a transcription factor varies widely: Some of them regulate only one single gene or operon, like the Lac repressor. These regulators are often present in rather low intracellular concentrations, e.g. 10-20 tetramers in case of the Lac repressor (Lin & Riggs, 1975; von Hippel *et al.*, 1974) Others, like MalT, regulate a small set of genes or operons (Schlegel et al., 2002), while others regulate a significant amount of genes as e.g. NarL that is - together with 6 other regulators - responsible for the control of 50 % of all genes in *E. coli* (Stewart, 1994).

Hitherto, CcpN could be characterised as a pure repressor controlling a small set of genes, and although efforts have been made to identify more targets of CcpN by microarray analysis and comparative transcriptome analysis, these attempts had remained unsuccessful (Servant *et al.*, 2005, Tännler *et al.*, 2008).

The aim of the present work was to identify potential new targets for CcpN. One promising new target, the *thyB* gene, has been investigated in detail. We demonstrate that CcpN is able to modestly activate transcription from the *thyB* promoter. Our results show that bioinformatics in combination with experimental methods is a powerful tool to identify new targets of transcriptional regulators.

5.3. Results and Discussion

CcpN is an abundant protein in *B. subtilis*

To determine the intracellular concentration of CcpN, protein crude extracts from *B. subtilis* DB104 in TY – together with purified CcpN of known concentration – were analysed by western blotting as described in Materials and Methods (Fig. 1). Since CcpN is constitutively expressed in log and stationary phase, cultures from $OD_{560} = 4.0$ were used. The amount of CcpN was calculated to be 4000 ± 600 molecules per cell. Taking into consideration a *B. subtilis* cell volume of 1×10^{-15} l (Abril *et al.*, 1997), the intracellular concentration of CcpN is approximately 6.6 µM.

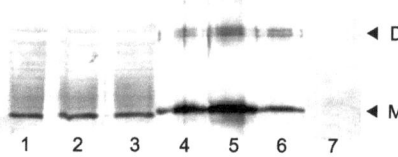

Figure 1: **Determination of the intracellular concentration of CcpN.**

Western Blot of a 12 % SDS polyacrylamide gel. 1-3: parallels of protein crude extract of *B. subtilis* DB104, corresponding to 1.2×10^8 cells; 4: 2.2 pmol of purified CcpN-His$_5$; 5: 4.4 pmol of purified CcpN-His$_5$; 6: 1.4 pmol of purified CcpN-His$_5$ in protein crude extract of DB104 (*ccpN::cat*); 7: 5 pmol BSA. The blot was developed using horseradish peroxidase catalysed conversion of diaminobenzidine, PC-BAS 2.08e software was used for quantification. M: band corresponding to the protein monomer; D: band corresponding to a putative protein dimer.

Intracellular amounts of transcriptional regulators vary from few molecules, e.g. 10-20 in case of the Lac repressor from *E. coli* (Lin & Riggs, 1975), to approximately 15000 like in the case of

CopR, a transcriptional repressor regulating the copy number of streptococcal plasmid pIP501 (Steinmetzer *et al.*, 1998). The pleiotropic regulators CcpA or CodY from *B. subtilis* are present in amounts resembling the one of CcpN, namely 3000 (Miwa *et al.*, 1994) and ≈2500 (A. L. Sonenshein, personal communication) molecules per cell, respectively. Intracellular repressor concentrations – in the case of chromosomally encoded repressors – appear to correlate at least partially with the amount of genes they regulate. The Lac repressor, present in very low concentration, regulates only one operon, while CcpA and CodY are involved in the direct regulation of 100 and 25 genes or operons, respectively (Sonenshein, 2007; Sonnenshein, 2005). Therefore, we wanted to find out if CcpN might regulate more than the three known genes *gapB*, *pckA* and *sr1*.

A database search reveals 291 potential CcpN targets

Therefore, the genome of *B. subtilis* was searched for possible CcpN binding sites using the SubtiList Web Server (http://genolist.pasteur.fr/SubtiList/) and a slightly revised version of the CcpN consensus sequence (TRTGHYATAYW) reflecting naturally occurring binding sites as well as binding sites found by EMSA (Licht *et al.*, 2005, Servant *et al.*, 2005). Additionally, one mismatch in the consensus sequence was allowed and only sequences within −100 bp or +20 bp relative to the translational start site were considered, since the known CcpN operators are located within this range (Servant *et al.*, 2005) and the location of the promoters of many genes is still unknown. 291 putative CcpN binding sites were found, among them 22 that perfectly matched the consensus sequence.

EMSA suggests at least three additional targets of CcpN

Out of all potential CcpN targets found, those encoding proteins involved in carbon catabolism or those whose CcpN operator sequence was matching the consensus binding site were selected for further investigation. EMSAs were performed using fragments carrying the putative CcpN operator in parallel with a fragment carrying the consensus binding site of the *sr1* promoter (Fig. 2). Of 20 investigated operators, only three were bound by CcpN: *thyB*, *gcaD* and *yhaM*. However, binding was less efficient than for the *sr1* operator. Apparent K_D values were 770 nM for *thyB*, 2.9 µM for *yhaM* and 3.4 µM for *gcaD*, compared to 420 nM for the *sr1* single site K_D value. Although the *thyB* operator shows one mismatch to the consensus sequence, its K_D value is still significantly higher than the K_D for *yhaM*, which is almost a perfect consensus sequence.

The *thyB* gene encodes the minor thymidylate synthase of *B. subtilis*, contributing to only 5 % of thymidylate synthesis (Neuhard *et al.*, 1978). Interestingly, the main thymidylate synthase of *B. subtilis* is closely related to thymidylate synthases encoded by phages, while *thyB* resembles the thymidylate synthases found in other bacteria (Tam & Borriss, 1998). The *yhaM* gene codes for a 5'-3' exoribonuclease (Oussenko *et al.*, 2002) and *gcaD* encodes UDP-N-acetylglucosamine pyrophosphorylase involved in cell-wall buildup (Hove-Jensen, 1992).

Since the K_D values for CcpN in the putative *gcaD* and *yhaM* operators indicated a very weak binding, we focused on *thyB* in further experiments.

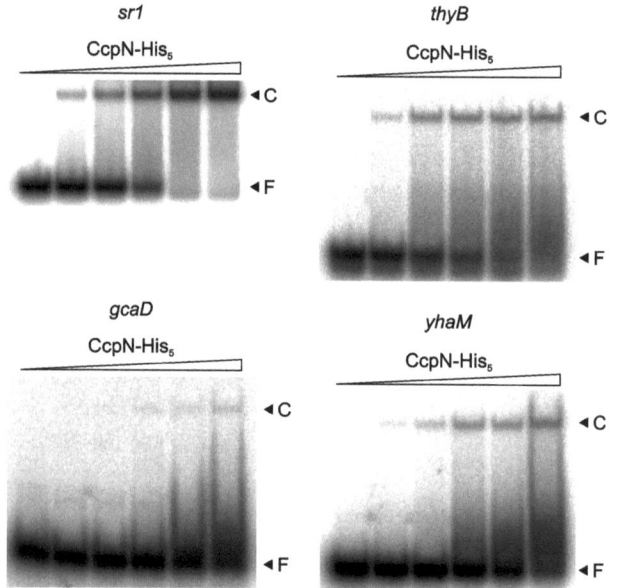

Figure 2: EMSA with different putative CcpN targets

EMSAs of double-stranded 23 bp DNA fragments containing the consensus binding site of the *sr1*, *gcaD*, *thyB* and *yhaM* operators. The DNA was incubated with increasing concentrations of purified CcpN-His$_5$ (CcpN concentration from left to right: 0 nM; 99 nM; 296 nM; 889 nM, 2.67 µM and 8 µM). The autoradiograms of the gels are shown.

DNase I footprinting reveals one large binding site at the *thyB* promoter

All previously investigated CcpN operators have two binding sites, one closely resembling the consensus binding site and one vastly differing from it (Licht & Brantl, 2006; Servant *et al.*, 2005). To determine binding sites at the *thyB* promoter, DNase I protection footprinting was performed (Fig. 3a). A situation resembling the *pckA* promoter was found, with only one long stretch of protected bases that includes both the –10 and the –35 region. As in *pckA*, the consensus binding site overlapped with the –10 region. The regions protected by CcpN are summarised in Figure 3a.

At the *sr1*, *pckA* and *gapB* promoters, there is one strong and one weak binding site, but these differences are overcome by CcpN binding cooperatively to the two sites, so that both sites are bound with the same efficiency when they are present on one DNA fragment and equally well protected from DNase I (Licht & Brantl, 2005). Interestingly, protection from DNase I was not constant within the protected region at the *thyB* promoter.

To verify that CcpN binds specifically at the *thyB* promoter, a *thyB* fragment carrying a mutation in the CcpN consensus binding site was subjected to DNase I footprinting. Indeed, no footprint was obtained (Fig. 3a).

Chemical interference footprinting identifies bases contacted by CcpN at the *thyB* promoter

Chemical interference footprinting experiments were performed to determine protein-DNA contacts at a higher resolution. Since neither contacts to C residues nor to the sugar-phosphate backbone played a significant role in the previously studied CcpN-operator interactions (Licht & Brantl, 2006) only methylation and $KMnO_4$ interference footprints were performed to detect G and A or T residues contacted by CcpN, respectively.

The *thyB* promoter shows a very unusual contact distribution compared with the previously investigated CcpN operators (Fig. 3b+c). The closest contacts are located within the consensus sequence, but the other contacts are almost evenly distributed over the remaining region protected in DNase I footprinting. A similar DNase I protection pattern was found at the *pckA* promoter, but chemical footprinting showed clearly two distinct binding sites, located two helical turns apart (Licht & Brantl, 2006). At the *thyB* promoter, three binding sites are present in total, one strong and two weaker ones. Since in DNase I footprints primarily the consensus sequence was protected, the other two binding sites might be auxiliary sites serving to guide CcpN to its main binding site – a feature that can also be observed with the Lac repressor (Oehler *et al.*, 1990). At the *thyB* promoter, contacts to T's are more prominent than contacts to G's and A's, whereas at the *sr1*, *gapB* and *pckA* promoters Gs were the bases forming the closest contacts with CcpN. Interestingly, the upstream binding site seems to rely mainly on contacts to A's whose modification occurs exclusively in the minor groove indicating that CcpN is able to access this much narrower DNA groove. Minor groove binding is often found in the form of 'indirect readout', like in CRP or HU of *E. coli* (Lindemose *et al.*, 2008, Swinger & Rice, 2007) and does not provide much binding specificity due to the low information content. However, it greatly increases binding stability. ComK, the *B. subtilis* competence regulator, was found to specifically bind its operator site through contacts in the minor groove, which leads to a novel potential activation mechanism (Smits *et al.*, 2007). It is possible that CcpN exerts the same mechanism at the *thyB* promoter.

A comparison of previously investigated CcpN operators and operator at the *thyB* promoter can be found in table S2.

Reporter gene assays and *in vitro* transcription demonstrate that CcpN acts as an activator at the *thyB* promoter

To assay the *in vivo* relevance of CcpN binding to the *thyB* promoter, transcriptional *thyB-lacZ* fusions were integrated into the chromosome and β-galactosidase activities measured. Strains were grown in glucose-free SP medium and glucose added where appropriate. The comparison between DB104 wild-type and isogenic *ccpN* knockout strain allowed for detection of an influence of CcpN. The results showed a small but significant and reproducible effect of CcpN (Fig. 4a). Whereas transcription is increased by a factor of 1.5 upon glucose addition in the wild-type strain, this effect cannot be observed in the *ccpN* knockout strain, where transcription levels in the absence and presence of glucose are nearly identical. To substantiate these results, a *lacZ*-fusion with a *thyB* promoter carrying a mutation that prevents CcpN-operator-interactions (see Fig. 3a) was

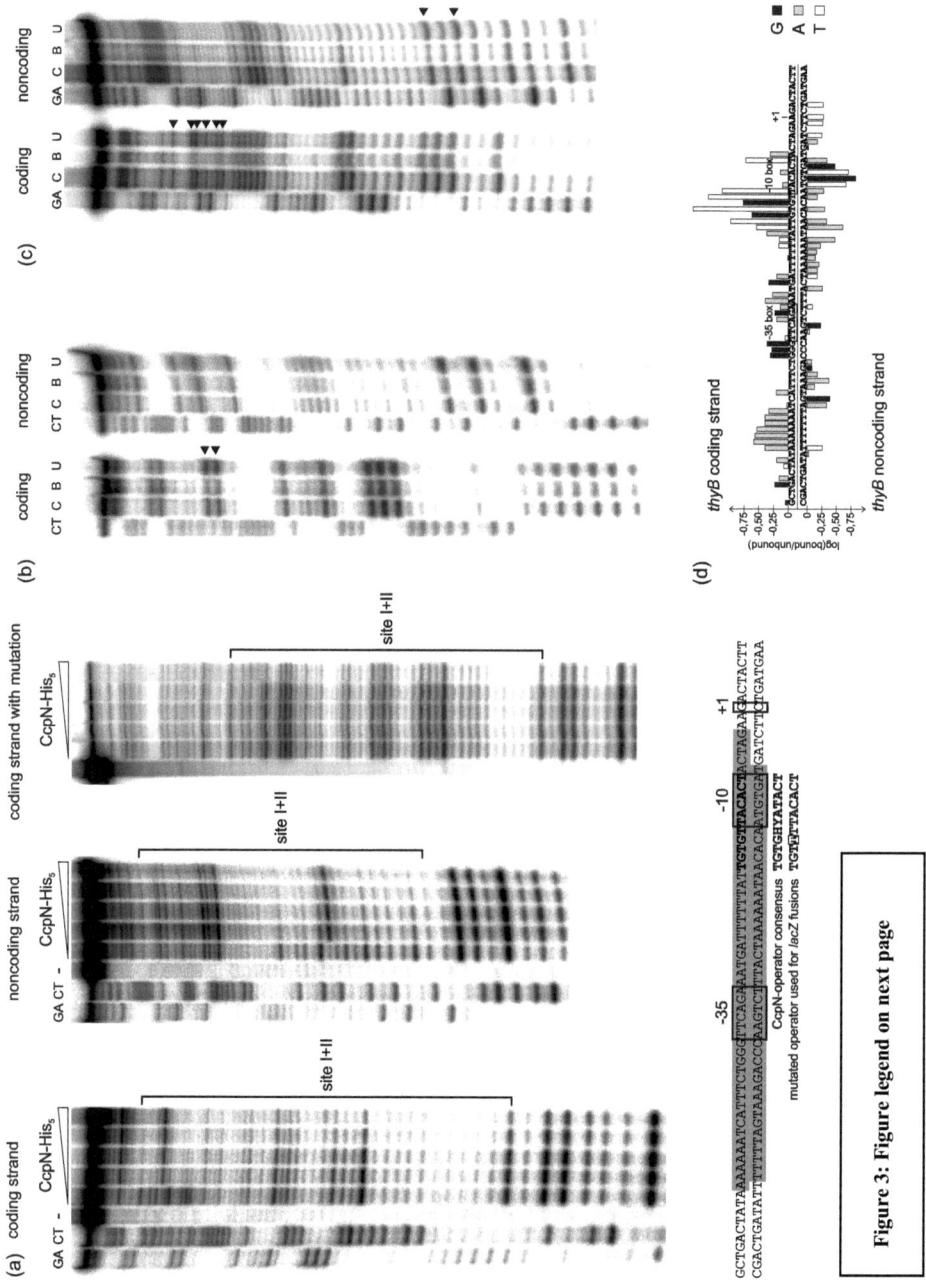

Figure 3: Figure legend on next page

Figure 3: DNase I footprinting and interference footprinting of the *thyB* promoter

(a) DNase I footprint. GA + CT: sequencing reaction; -: control without DNase I. The DNA was incubated with increasing concentrations of purified CcpN-His$_5$ (CcpN concentration from left to right: 0 nM; 296 nM; 889 nM, 2.67 µM and 8 µM) prior to DNase I treatment. The combined protected sites have been designated site I+II. The autoradiograms of the gels are shown. Left, wild-type *thyB* promoter; right, mutated *thyB* promoter. An Overview of the protected region at the *thyB* promoter is shown below the gels. Bases protected by CcpN are coloured with a grey background. –35 and –10 regions, the transcriptional start site, the CcpN-operator consensus and the mutated operator sequence used in *lacZ* fusions (mutation shown in inverted colours) are indicated.

(b) Methylation interference footprint of the *thyB* promoter. CT: Maxam-Gilbert C+T sequencing reaction; C, control (protein-free methylated DNA); B and U, bound and unbound fraction of methylated DNA subjected to binding with CcpN-His$_5$. Close contacts are indicated by black triangles.

(c) KMnO$_4$ interference footprint of the *thyB* promoter. Abbreviations are as in C. GA: Maxam-Gilbert G>A sequencing reaction; C, control (protein-free KMnO$_4$-treated DNA).

(d) Column diagrams indicating the relative strength of interference signals for both strands of the three operators. Only positive signals, i.e. signals that indicate contacts, are shown. Measured values are averaged from four independent experiments.

constructed and analysed as above. The site and type of mutation was chosen based on previously performed EMSAs (Licht *et al.*, 2005), where the mutated base has been shown to be invariant. This construct did not respond to glucose, indicating that CcpN truly regulates the *thyB* promoter (Fig. 4a). This finding is particularly striking, since CcpN has only been shown to work as a repressor under glycolytic conditions (Licht *et al.*, 2005, Servant *et al.*, 2005), but here, it acts as an activator. A statistical significance test confirmed the difference in activation between wild-type and knockout strain with a confidence of > 98 %. Compared to the huge effects CcpN exerts as a repressor, the relatively small effect observed here might be due to the significantly greater K_D value observed for the *thyB* promoter. Beside CcpN, there are few other examples of DNA-binding proteins that use an identical consensus sequence for both activation and repression. Two prominent examples are the global regulators CodY and CcpA from *B. subtilis* (Henkin, 1996; Sonenshein, 2005), the former clearly resembling the case of CcpN, whereas the latter requires HPr as a co-regulator and acts as activator or repressor depending on operator positions.

To corroborate these findings, *in vitro* transcription experiments with *B. subtilis* RNA polymerase were performed (Fig. 4b). This assay clearly showed that CcpN is able to activate *thyB* transcription under conditions of low pH and high ATP concentration, conditions found to be required for CcpN at the three known target promoters. This is in good agreement with the results of the β-galactosidase measurements, showing that CcpN is able to activate the transcription of *thyB* both *in vivo* and *in vitro*. It remains unclear how CcpN is able to act as an activator upon binding to the -10 region, but other studies have shown the same effects for *B. subtilis* CodY at the *ackA* promoter and the *E. coli* MerR protein (Shivers *et al.*, 2006; O'Halloran *et al.*, 1989).

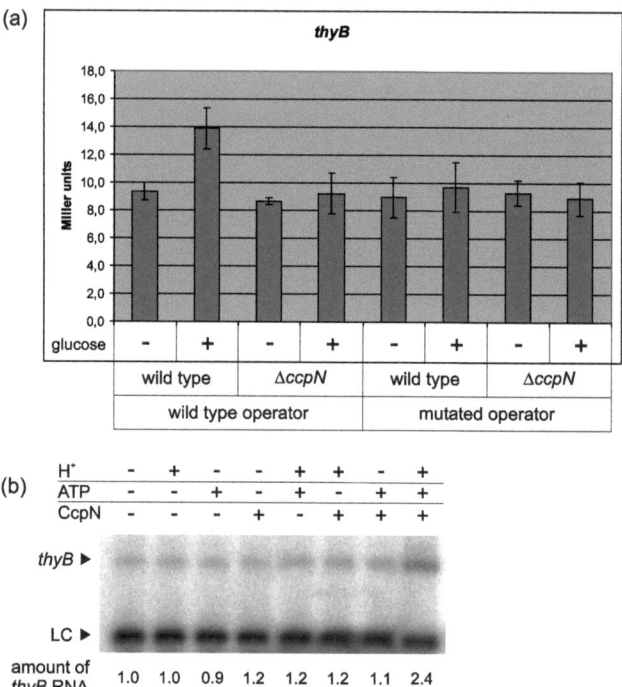

Figure 4: β-galactosidase assays and *in vitro* transcription at the *thyB* promoter

(a) Column diagrams showing the amount of Miller units measured at the *thyB* (with wild type or mutated CcpN operator) promoter under repressing and non-repressing conditions in the DB104 wild-type and *ccpN* knockout strains. The cells were grown in SP medium till an OD_{600} of 2.0. Values are averaged over 6 different clones and three independent experiments.

(b) *In vitro* transcription assay using 100 nM of a DNA fragment and 50 nM of His-tagged *B. subtilis* RNAP as well as 50 nM purified SigA in each reaction. The autoradiogram of the gel is shown. Reaction conditions are denoted above each lane and the *thyB* transcript is indicated by a black arrow. One of three independent experiments is shown. LC, loading control.

Regarding the physiological relevance of CcpN-mediated activation of *thyB transcription*, one could hypothesise that more thymidylate is needed during glycolysis, which usually comes along with excellent growth conditions and increased growth rate. ThyB contributes to only about 5 % of TSase activity in *B. subtilis* (Neuhard *et al.*, 1978), and an activity increase of 50 % does not seem relevant in the background of ThyA. However, ThyB might be the "original" TSase of *B. subtilis'* ancestors, and while its activity has been almost lost, the regulatory mechanism is still intact. The results presented here might also indicate that ThyB is involved in the formation of pyrimidine nucleotide activated sugars during glycolysis and therefore upregulated independently of ThyA.

Since it is not excluded that CcpN might regulate thymidylate synthetase genes in other Gram-positive bacteria that possess – in contrast to *B. subtilis* - only one *thy* gene, an alignment of CcpN binding sites at *thyB* promoters in related Gram-positive species was performed.

Alignments show potential CcpN binding sites at *thyB* promoters in related Gram-positive species

To investigate if CcpN operator sites are located upstream of *thyB* genes in these species, all bacteria encoding CcpN homologues were scanned for *thyB* related genes using BLAST (Zhang et al., 2000). Species with the highest *thyB* homology compared to *B. subtilis* were subsequently searched for CcpN operator sites upstream of the *thyB* start codon, since the transcriptional start sites are not known in most of these organisms. Figure 5 shows the alignments. Interestingly, in *B. amyloliquefaciens*, *B. anthracis*, *B. cereus* and *S. aureus* as well as in *F. nucleatum* consensus-like CcpN operators are located upstream of the *thyB* genes in approximately the same distance as in *B. subtilis*. In other bacteria, more mismatches were found in the operator sequence and operator spacing compared to *B. subtilis*. However, it is conceivable that other organisms have other requirements for CcpN consensus operator sites, and that these sites bind their corresponding CcpN proteins quite well.

```
B. subtilis          AT TGTGTTAC ACT ACTAGAA GACTACTTTT AAAGGATGAA AAAA ATG
B. amyloliquefaciens AT TGTGTTAC ACT ACTAAAA GACTACTTTA AAAGGATGAA AAAA ATG
B. anthracis         GT TATGATAA ATT GCATATA AGATTGAAAG AAGGTTTTAC TACAT ATG
B. cereus            TT TATGATAA AAT CCATATA GACTAGAGAT TTGAAAGAAG GTTTTCTAC AT ATG
B. halodurans        TGTGCTAGTG AGGATACATA AGCAGATGG AT TGTGCATA ATA TGTACTA
                     AATGACAATC CTCCATGCAT ATTTTTCGTT TTCGCCGGGT GAACGAAGTC
                     GCACTGGCGA AAAGGATCAA ATG
F. nucleatum         TA TGTGGTAT AAT AATATAA AATTAAAAAA ATAGGAGAGA GGAAA ATG
O. iheyensis         TT TATGTTCT AAT AGTGAGT TTATGTTTTA AGCTTATGTA ACAAAAGAGG
                     ACTAGTA TCT GTTATAAT TA AATCATTTGA GAAGCAAAAT CTATATTTTA
                     AATATAGTCA GGAGAGATTA ATAT ATG
S. aureus            AT TATGTCGA ACT AAATATA CATACTATAA ATAATGAAAA TGAGGTGTTA TCGCAT ATG
S. epidermitis       TA TATGAGAT ATT TTACATA AAAGGTTTAA AATTTTATAA ACTCAAAACC
                     CTCTCTCTTC TATCGATACA TCTTTTCTAG TATAATGTTT ATTAATATAA
                     TACATATTTT TATAAGGAGT GTGTACGC ATG
S. agalactiae        TT TTTGCTAA ATT CCATTTT ACCATAAAGA AGCTAAAAAT ATGAAAAAAA
                     GCTTTAACCT TCAAAGTCTT GTTTCTCACT AGAATATCTT TTTAAAA TCT
                     GATAAAAT AA GACTTTGGAG GTATCTC ATG
```

Figure 5: Alignment of promoter regions

Alignment showing the regions upstream of the translational start site of *thyB* of different bacteria possessing a *ccpN* homologue or orthologue. The translational start site is indicated in bold and underlined. Sequences resembling the *B. subtilis* consensus binding site for CcpN have been labelled in bold and with black rectangles. Deviations from the consensus sequence are marked by grey letters.

Overall, it appears that CcpN is involved in the regulation of one more *B. subtilis* gene than previously anticipated and upcoming searches might reveal more new targets. At least the high degree of conservation regarding binding sites, even between different organisms, strongly favours this hypothesis.

This report demonstrates that bioinformatics in concert with molecular biological methods can be used to identify new targets of a transcriptional regulator, even if the biological context in which this regulator acts on these targets is not yet understood. In the case of *thyB*, analyses in other firmicutes that have both a CcpN homologue and a CcpN binding site upstream of their *thyB* genes and encode – in contrast to *B. subtilis* – only one thymidylate synthetase, might shed light on a possible role of CcpN-mediated regulation of these genes.

5.4. Materials and Methods

Enzymes and chemicals

Chemicals used were of the highest purity available. *E. coli* RNA polymerase and all chemicals were purchased from Sigma-Aldrich™. Taq-polymerases were purchased from Roche (Germany) and Solis Biodyne (Estonia).

Strains, media and growth conditions

B. subtilis strains DB104 (Kawamura & Doi, 1984) and DB104 (*ccpN::cat*) (Licht *et al.*, 2005) were used. TY medium (Licht *et al.*, 2005) served as complex medium. SP medium (Preis *et al.*, in press) served as glucose-free medium. Strain NIG2001 was used for expression of His-tagged *B. subtilis* RNA polymerase (Fujita & Sadaie, 1998b). *E. coli* strain DH5α was used for cloning and strain BL21 (DE3) (pETSigA) for overexpression and purification of His-tagged *B. subtilis* SigA (Fujita & Sadaie, 1998a).

Overexpression and purification of proteins

CcpN overexpression and purification were performed as published before (Licht *et al.*, 2008). Expression and purification of His-tagged *B. subtilis* RNA polymerase and His-tagged SigA were carried out as described (Fujita & Sadaie, 1998a, 1998b).

Determination of the intracellular concentration of CcpN in *B. subtilis*

The intracellular concentration of CcpN was determined following the procedure described for CopR (Steinmetzer *et al.*, 1998) except that the Western blot was developed with diaminobenzidine.

Table 1: Strains and plasmids used in this study

Strain	Genotype	Reference
E. coli K12 DH5α	F- φ80lacZΔM15 Δ(lacZYA-argF) U169 recA1 endA1 hsdR17 (r_k^-, m_k^+) phoA supE44 λ-thi-1 gyrA96 relA1	Invitrogen™
E. coli BL21 (DE3) (pETSigA)	dcm ompT hsdS(rB-mB-) gal λ(DE3)	Fujita et al., 1998
B. subtilis NIG2001	trpC2 pheA1 neo^r rpoC-His$_6$	Fujita et al., 1998
B. subtilis DB104	his nprR2 nprE18 ΔaprA3	Kawamura & Doi, 1984
B. subtilis DB104	his nprR2 nprE18 ΔaprA3 (ccpN::cat)	Licht et al., 2005

Plasmid	Description	Reference
pAC6	pBR322 based vector for integration of transcriptional lacZ fusions into amyE locus of B. subtilis, ApR, CmR	Stülke et al., 1997
pTHY1	pAC6 with pthyB-lacZ fusion	this work
pATM1	pAC6 with pthyB-lacZ fusion with mutated CcpN operator	this work

EMSAs and footprinting experiments

EMSAs, methylation and potassium permanganate interference footprinting were performed as described (Licht & Brantl, 2006). DNase I footprinting was performed as described (Licht et al., 2005).

Construction of plasmids for transcriptional *lacZ* fusions and measurements of β-galactosidase activities

Plasmid pAC6 was used to insert an EcoRI-BamHI fragment obtained by PCR from chromosomal DNA of B. subtilis with oligodeoxyribonucleotides SB1069 (Table S1) and SB1070 yielding plasmid pTHY1. For plasmid pATM2, oligodeoxyribonucleotides SB1268 and SB1069 as well as SB1267 and SB1070 were used on chromosomal DNA of B. subtilis DB104 as template to create fragments MUT2up and MUT2down, respectively. A second PCR using these fragments and oligodeoxyribonucleotides SB1069 and SB1070 resulted in fragment MUT2, carrying the thyB promoter and a mutated CcpN operator site, which was inserted as an EcoRI-BamHI fragment into plasmid pAC6. Integration of the plasmids into the amyE locus and measurements of β-galactosidase activities were performed as described previously (Brantl, 1994).

***In vitro* transcription**

In vitro transcription reactions were performed in a final volume of 10 μl in in vitro transcription buffer (40 mM Tris/acetate, pH 7.5, 10 mM magnesium acetate, 100 mM potassium acetate and 20% glycerol) in the presence of 1 mM GTP, 0.1 mM ATP, 0.1 mM CTP, 0.01 mM

UTP and 0.011 µM [α-^{32}P]UTP with templates generated as described (Licht et al., 2008). If appropriate, effectors were added, followed by the addition of 100 nM of double-stranded DNA template and 50 nM of His-tagged B. subtilis RNA polymerase and 50 nM SigA-His$_6$ and incubation for 30 min at 37 °C. Samples were treated with formamide loading dye and separated on a 6 % denaturing polyacrylamide gel at 300 V/25 mA for 50 min. Dried gels were subjected to PhosphorImaging as above.

Acknowledgements

We thank M. Fujita for the strains for expression of His-tagged B. subtilis RNA polymerase and sigma factor. Furthermore, we thank C. Wiedemann for helping with the measurements of the lacZ fusions. This work was supported by grant BR1552/6-2 from Deutsche Forschungsgemeinschaft (to S. B.). A. L. is financed by a scholarship from the federal state of Thuringia and from Deutsche Forschungsgemeinschaft.

5.5. References

1. Abril AM, Salas M, Andreu JM, Hermoso JM & Rivas G (1997) Phage Φ29 protein p6 is in a monomer-dimer equilibrium that shifts to higher association states at the millimolar concentrations found in vivo. Biochemistry 36: 11901-11908.

2. Brantl S (1994) The copR gene product of plasmid pIP501 acts as a transcriptional repressor at the essential repR promoter. Mol Microbiol 14: 473-483.

3. Chambliss G H (1993) Carbon source mediated catabolite repression. In Bacillus subtilis and other gram-positive bacteria: biochemistry, physiology, and molecular genetics (Sonenshein, A. L., Hoch, J. A. & Losick, R., eds), pp. 212-219, Am. Soc. Microbiol., Washington, DC.

4. Fujita M & Sadaie Y (1998) Promoter selectivity of the Bacillus subtilis RNA polymerase sigmaA and sigmaH holoenzymes. J Biochem 124: 89-97.

5. Fujita M & Sadaie Y (1998) Rapid isolation of RNA polymerase from sporulating cells of Bacillus subtilis. Gene 221: 185-90.

6. Henkin TM (1996) The role of CcpA transcriptional regulator in carbon metabolism in Bacillus subtilis. FEMS Microbiol Lett 135: 9-15.

7. Hove-Jensen B (1992) Identification of tms-26 as an allele of the gcaD gene, which encodes N-acetylglucosamine 1-phosphate uridyltransferase in Bacillus subtilis. J Bacteriol 174: 6852-6856.

8. Kawamura F & Doi RH (1984) Construction of a Bacillus subtilis double mutant deficient in extracellular alkaline and neutral proteases. J Bacteriol 160: 442-444.

9. Lewis M (2005) The lac repressor. C R Biol 328: 521-548.

10. Licht A & Brantl S (2006) Transcriptional repressor CcpN from Bacillus subtilis compensates asymmetric contact distribution by cooperative binding. J Mol Biol 364: 434-448.

11. Licht A., Preis S & Brantl S (2005) Implication of CcpN in the regulation of a novel untranslated RNA (SR1) in *Bacillus subtilis*. *Mol Microbiol* 58: 189-206.

12. Licht A, Golbik R and Brantl S (2008) Identification of ligands affecting the activity of the transcriptional repressor CcpN from *Bacillus subtilis*. *J Mol Biol* 380: 17-30.

13. Lin S & Riggs AD (1975) The general affinity of *lac* repressor for *E. coli* DNA: implications for gene regulation in procaryotes and eucaryotes. *Cell* 4: 107-111.

14. Lindemose S, Nielsen PE & Møllegaard NE (2008) Dissecting direct and indirect readout of cAMP receptor protein DNA binding using an inosine and 2,6-diaminopurine *in vitro* selection system. *Nucleic Acids Res* 36: 4797-4807.

15. Maas, WK (1994) The arginine repressor of *Escherichia coli*. *Microbiol Rev* 58:631-640.

16. Miwa Y, Saikawa M & Fujita Y (1994) Possible function and some properties of the CcpA protein of *Bacillus subtilis*. *Microbiology* 140: 2567-2575.

17. Neuhard J, Price AR, Schack L & Thomassen E (1978) Two thymidylate synthetases in *Bacillus subtilis*. *Proc Natl Acad Sci USA* 75: 1194-1198.

18. Oehler S, Eismann ER, Krämer H & Müller-Hill B (1990) The three operators of the lac operon cooperate in repression. *EMBO J* 9: 973-979.

19. O'Halloran TV, Frantz B, Shin MK, Ralston DM & Wright JG (1989) The MerR heavy metal receptor mediates positive activation in a topologically novel transcription complex. *Cell* 56: 119-129.

20. Oussenko IA, Sanchez R & Bechhofer DH (2002) *Bacillus subtilis* YhaM, a member of a new family of 3'-to-5' exonucleases in gram-positive bacteria. *J Bacteriol* 184: 6250-6259.

21. Preis H, Eckart RA, Gudipati RK, Heidrich N & Brantl S (2009) CodY activates the transcription of a small RNA in *Bacillus subtilis*. *J Bacteriol* in press.

22. Schlegel A, Böhm A, Lee SJ, Peist R, Decker K & Boos W (2002) Network regulation of the *Escherichia coli* maltose system. *J Mol Microbiol Biotechnol* 4: 301-307.

23. Servant P, Le Coq D, & Aymerich S (2005) CcpN (YqzB), a novel regulator for CcpA-independent catabolite repression of *Bacillus subtilis* gluconeogenic genes. *Mol Microbiol* 55: 1435-1451.

24. Shivers RP, Dineen SS & Sonenshein AL (2006) Positive regulation of *Bacillus subtilis ackA* by CodY and CcpA: establishing a potential hierarchy in carbon flow. *Mol Microbiol* 62: 811-822.

25. Smits WK, Hoa TT, Hamoen LW, Kuipers OP & Dubnau D (2007) Antirepression as a second mechanism of transcriptional activation by a minor groove binding protein. *Mol Microbiol* 64: 368-381.

26. Sonenshein AL (2005) CodY, a global regulator of stationary phase and virulence in Gram-positive bacteria. *Curr Opin Microbiol* 8: 203-207

27. Sonenshein AL (2007) Control of key metabolic intersections in *Bacillus subtilis*. *Nat Rev Microbiol* 5: 917-927.

28. Steinmetz M (1993) Carbohydrate catabolism: pathways, enzymes, genetic regulation, and evolution. In *Bacillus subtilis and other gram-positive bacteria: biochemistry, physiology, and molecular genetics* (Sonenshein, A. L., Hoch, J. A. & Losick, R., eds), pp. 157-170, Am. Soc. Microbiol., Washington, DC.

29. Steinmetzer K, Behlke J & Brantl S (1998) Plasmid pIP501 encoded transcriptional repressor CopR binds to its target DNA as a dimer. *J Mol Biol* 283: 595-603.

30. Stewart V (1994) Dual interacting two-component regulatory systems mediate nitrate- and nitrite-regulated gene expression in *Escherichia coli*. *Res Microbiol* 145: 450-454.

31. Stülke J & Hillen W (2000) Regulation of carbon catabolism in Bacillus species. *Annu Rev Microbiol* 54: 849-880.

32. Stülke J, Martin-Verstraete I, Zagorec M, Rose M, Klier A & Rapoport G (1997) Induction of the *Bacillus subtilis ptsGHI* operon by glucose is controlled by a novel antiterminator, GlcT. *Mol Microbiol* 25: 65-78.

33. Swinger KK & and Rice PA (2007) Structure-based analysis of HU-DNA binding. *J Mol Biol* 365: 1005-1016

34. Tam NH & Borriss R. (1998) Genes encoding thymidylate synthases A and B in the genus *Bacillus* are members of two distinct families. *Mol Gen Genet* 258: 427-430.

35. Tännler S, Fischer E, Le Coq D, Doan T, Jamet E, Sauer U & Aymerich S (2008) CcpN controls central carbon fluxes in *Bacillus subtilis*. *J Bacteriol* 190: 6178-6187.

36. von Hippel PH, Revzin A, Gross CA & Wang AC (1974) Non-specific DNA binding of genome regulating proteins as a biological control mechanism: I. The lac operon: equilibrium aspects. *Proc Natl Acad Sci USA* 71: 4808-4812.

37. Yamada M, Makino M, Amemura M, Shinagawa H & Nakata A (1989) Regulation of the phosphate regulon of *Escherichia coli*: analysis of mutant *phoB* and *phoR* genes causing different phenotypes. *J Bacteriol* 171: 5601-5606.

38. Zhang Z, Schwartz S, Wagner L & Miller W (2000) A greedy algorithm for aligning DNA sequences. *J Comput Biol* 7: 203-214.

6. Summary

The scope of this work was the detailed characterisation of the transcription factor CcpN of the Gram-positive bacterium *B. subtilis* which regulates the genes *sr1*, *pckA* and *gapB*.

To this end, bases within the operators that are necessary for forming contacts with CcpN were determined at the *sr1*, *pckA* and *gapB* promoters by interference footprinting. These experiments showed that intensive contacts were made within a sequence corresponding to the previously determined consensus sequence and that each promoter consists of two operator segments with different contact strength. EMSAs demonstrated that these differences in contact strength also resulted in a varying binding strength of the single operators when present on a separate DNA fragments. However, if both operators were located on one DNA fragment, both were bound with equal strength and in addition more intensively than the single operator sites. Regarding the change in shape and slope of the binding curves when comparing single operators with operator pairs and the energy gain resulting from CcpN-operator interaction it was concluded that CcpN binds its operator sites cooperatively. Furthermore, energetic calculations of CcpN-DNA interaction at different temperatures revealed that the binding process is driven by a strong enthalpy rather than strong entropy, ensuring a stable interaction of CcpN with its operators over a large temperature scale.

Since CcpN is present in constant concentrations within the cell and the expression of its gene was found to be not regulated, a search for intracellular effectors of CcpN was performed. *In vitro* transcription reactions showed that purified CcpN without an effector is able to specifically repress transcription at the three investigated promoters and provides an explanation for the strong derepression observed in a *ccpN* knockout strain. ATP and low pH were identified as the intracellular activators of CcpN activity, fitting quite well into CcpN's scheme of action: During glycolysis, when CcpN is active, ATP levels in the cell are high and the cytosol becomes slightly acidic due to acetate production. On the contrary, ADP has been shown to counteract the activating effect of ATP at equimolar concentrations. Both results have been substantiated by CD spectroscopy, which showed extensive structural rearrangements of CcpN upon ATP binding at low pH, but not at neutral pH, while ADP binding did only result in weak structural alterations. It was thus concluded that the repression activity of CcpN is stimulated by structural alterations induced by ATP binding at low pH and repressed by ADP binding.

Two CBS domains, which are able to bind adenosine residues, have been found within the CcpN sequence. To elucidate the role of these domains, a CcpN mutant with an amino acid exchange in a conserved residue in one domain was investigated. This mutant, although still being able to bind to DNA and showing the same structure as the wild type, did no longer respond to ATP, indicating that the CBS domains are indeed responsible for ATP binding.

Eventually, EMSAs performed under repressing and non-repressing conditions showed that the positive effectors of CcpN did not alter the affinity to its operators.

Since nothing was known about the repression mechanism of CcpN, efforts have been made to elucidate this mechanism at the three investigated promoters. EMSAs have demonstrated that

CcpN and RNAP are able to bind together to the promoter under repressive and non-repressive conditions. However, competition for promoter binding between these two proteins has been observed which explains the strong derepression in a *ccpN* knockout strain. Using open complex formation assays it has been demonstrated that open complexes can still be formed under repressive conditions at all three promoters, showing that CcpN does not prevent melting of the DNA. At the *gapB* promoter, no abortive transcripts were detectable under repressive conditions indicating that CcpN represses transcription initiation at this promoter. The *sr1* and *pckA* promoters still showed abortive transcript synthesis under repressive conditions, but the transition to the elongation complex was inhibited.

Far western blot and co-elution interaction studies showed that CcpN is able to specifically interact with the α-subunit of RNAP, but not with the σ-factor. This suggests that the repression at the *sr1* and *pckA* promoters occurs by interaction between CcpN and the α-subunit, which in turn stalls the RNAP at the promoter. This mechanism can, however, be excluded for the *gapB* promoter, since the operator location does not allow CcpN to be positioned in a way to properly contact the α-subunit. Here, an alteration in DNA structure upon CcpN binding has been detected in preceding investigations, which is potentially responsible for the prevention of transcription initiation.

Early characterisations of CcpN already suggested a rather high intracellular concentration. This observation was substantiated using western blots, and the amount of CcpN was determined to be 4000 molecules per cell. Since there is a certain correlation between repressor concentration and the number of regulated genes, a search for new CcpN target genes was performed in the genome of *B. subtilis*. The search resulted in numerous potential operators, but only those located in the promoter region of the *thyB*, *yhaM* and *gcaD* genes showed CcpN binding *in vivo*, with *thyB* being the only one strong enough to be investigated further. Interestingly, the *thyB* CcpN operator showed a binding pattern not observed before, where one strong and two weak operators are conjoined without a spacer region.

Using *lacZ* transcriptional fusions and *in vitro* transcription, a slight activation of the *thyB* promoter by CcpN under glycolytic conditions has been shown. This regulation seems to be of little physiological relevance in *B. subtilis*, since the *thyB* gene is hardly active, and thymidylate synthase activity is mainly carried out by ThyA. However, related species containing only a *thyB* and no *thyA* gene have CcpN operator sequences upstream of their respective *thyB* genes and it is feasible that CcpN plays an important regulatory role in *thyB* expression in these bacteria.

The results of the presented study greatly increase our understanding of the transcription factor CcpN from *B. subtilis*. This makes – together with preceding works – CcpN to one of the best characterised transcription factors in *B. subtilis* including one of the few for which a detailed working model (summarised in Figure 1) and a repression mechanism has been determined.

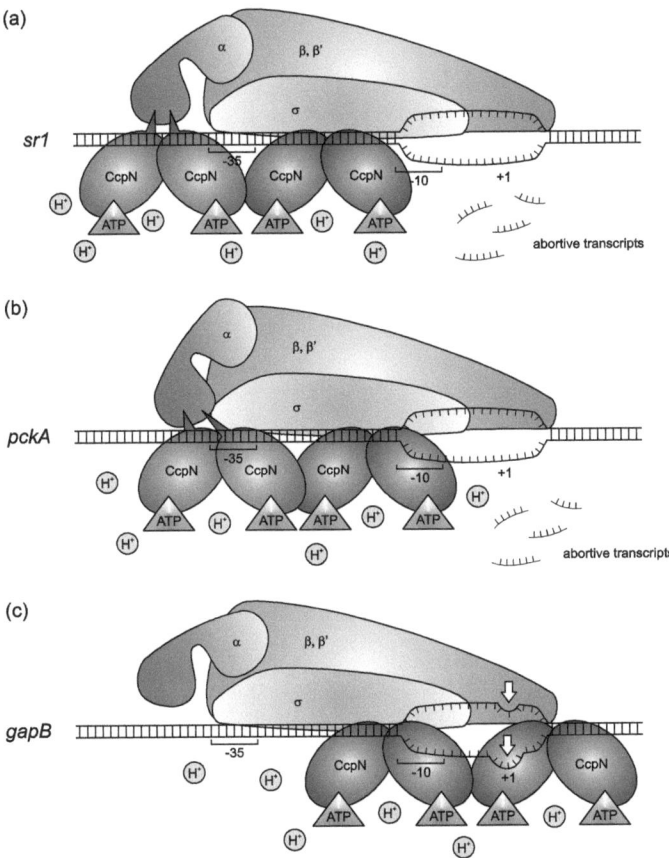

Figure 1: Current model of CcpN activity under glycolytic condition

(a) Situation at the *sr1* promoter. CcpN and the RNAP are bound to the promoter. Due to glycolysis, the ATP level of the cell is high while the ADP level is low, allowing ATP to bind to CcpN. In addition, intracellular pH is lowered by accumulating acetate. CcpN can interact with the α-subunit of RNAP, still allowing the formation of abortive transcripts, but not the transformation into an elongation complex. The "spikes" protruding from CcpN symbolise the interaction with the α-CTD.

(b) Situation at the *pckA* promoter. The situation is, apart from slightly different operator positions, similar to the *sr1* promoter.

(c) Situation at the *gapB* promoter. Because of the operator positions, CcpN cannot contact the α-subunit. It is conceivable that CcpN induces structural changes in the DNA instead, which ultimately inhibit the synthesis of abortive transcripts. White arrows denote positions, where changes in DNA structure that could interfere with transcription have been detected. Alternatively, a direct interaction between CcpN and the β or β' subunit of the RNAP is possible.

Die VDM Verlagsservicegesellschaft sucht für wissenschaftliche Verlage abgeschlossene und herausragende

Dissertationen, Habilitationen, Diplomarbeiten, Master Theses, Magisterarbeiten usw.

für die kostenlose Publikation als Fachbuch.

Sie verfügen über eine Arbeit, die hohen inhaltlichen und formalen Ansprüchen genügt, und haben Interesse an einer honorarvergüteten Publikation?

Dann senden Sie bitte erste Informationen über sich und Ihre Arbeit per Email an *info@vdm-vsg.de*.

Sie erhalten kurzfristig unser Feedback!

VDM Verlagsservicegesellschaft mbH
Dudweiler Landstr. 99   Telefon +49 681 3720 174
D - 66123 Saarbrücken   Fax   +49 681 3720 1749
www.vdm-vsg.de

Die VDM Verlagsservicegesellschaft mbH vertritt

Printed by Books on Demand GmbH, Norderstedt / Germany